Table of contents

The bed of nails..
In the Moscow Institute of Experimental surgery ....
Kuznetsov's story... 4
As crazy as idea seemed -- it worked! 6
Thousands of tests in clinics.. 7
Medical Newspaper... 7
Insomnia, colds and flu, asthma, and panic attacks........................ 8
During the 1991 coup in Moscow... 8
Kuznetsov's Team... 10
Arthritis, Leukemia, or Cerebral Palsy. 12
What Do They Have in Common?.. 12
"Very simple: it's a panacea" 12
Chernobyl's babies.. 13
Radiation therapy... 13
Sarcoma... 14
Hypotheses Needed.. 14
Self-acupuncture?.. 16
Can acupoints explain the effects?.. 16
Self-diagnosis, Neurophysiology, and Neuro-computers.................... 17
Skin, spinal cord, and sympathetic nervous system....................... 18
The skin projections of internal organs 20
(zones of Zakharian-Head) 20
Why Do Children Rub Their Bruises? 21
The Largest Organ of Your Body May Be Starving......................... 23
Body awareness... 24
Research: the Pilot Study.. 24
The Panacea mat used in our pilot study in 1995......................... 25
Results... 25
Effects of applicator... 27
Major Practical Application.. 27
STRESS.. 28
EXERCISE.. 29
WEIGHT LOSS... 30
MEN'S HEALTH: 31
An Interesting Case of Cured Impotence..................................... 31
WOMEN'S HEALTH (PMS and hot flashes)....................................... 31
Case histories.. 33
Testimonials of Users of the Panacea Concerning 37
Its Effectiveness... 37
Summaries of protocols of clinical trials *39*
Disclaimer... *42*
APPENDIX... *42*
Physiological effects of endogenous endorphins.......................... 42
   The Releasers... 43
      Endorphin-induced release of regulatory peptides. 43
      Adaptatogenesis.. 46

How Endorphins Control Body Functions............ 47
    Positive reward 47
    Placebo 47
    Stress 47
    Appetite/Hunger 47
    Body temperature 48
    Blood glucose 48
    Immune system 48
    Growth hormone 48
    Other hormones 49
    Oxygen utilization 49
  The role endorphins in diseases 49
  PLICKATOR in the Health Journal............ 51
  IPLICKATOR against insomnia 53
  PLICKATOR helps arthrosis patients............ 55

The bed of nails in Sweden............ 58
  Shaktimattan is registered in the Guiness Book of Records............ 58
  Comparing Kuznetsov's Applicator with "Shakti mat" 60
  New version of applicator origine: "I wandered down from the mountain and created a bed of nails"............ 60
  References............ 61

The great success of the "Kuznetsov's IPLIKATOR" as a self-help healing devise never ceased to amaze me. The word "IPLIKATOR"

The inventor, Ivan I. Kuznetsov

is an acronym of Russian words meaning in English: "Needle-like, Prophylactic, Healing, Activating, Toning, Providing Efficiency." The Russians started using the word "applicator" which has the same meaning as in English. This device could best be described as "the bed of nails."

The bed of nails

The ancient Yogi's idea of "the bed of nails" first came to Russia in the mid-19th century thanks to the celebrity Ivan Turgenev and his sensational novel "The Eve." The hero, a Bulgarian revolutionary

Insarov, would lie down on nails to challenge his spirit. Thus, it became a symbol of revolutionary romanticism, and was very attractive to liberal youth and quite repulsive to the conservative establishment.

It would be wrong to say that the progress of the Applicator was met with the classic resistance of the stagnation period prevalent during Kuznetsov's time: It did not escape the attention of five Moscow clinics, which gave it positive reviews. The Department of Health also gave approval for the IPLIKATOR. It was mass produced and distributed by Kuznetsov's cooperative firm, and could be bought in any drug store. 70 million were bought before a manufacturing collapsed in the 1990s! Now it's back and sold all over the world, along with the Swedish version, the Shakti mat, which was a real, making headlines sensation and even registered in the Guinness book of records.

The case files of the patients took up walls of space in Kuznetsov's office. The line for appointments formed well before opening the doors. A documentary was filmed and aired through the national television. But, at the same time, there existed virtually no ideas on exactly how the Applicator worked, as there were no existing professional publications whatsoever.

In the Moscow Institute of Experimental surgery
It was the Moscow Institute of Experimental surgery where I first heard about the "Applicator of Kuznetsov". My husband worked for the institute and one day came home very excited about a new method for pre-op treatment of patients resistant to regular medicine, or allergic to it, etc. "Just imagine a rubber mat pierced

with needles and a patient has to lie down on it and to stay for an hour or more. They say, it never pierces the skin and hurts for only minute or two, then people get warmed and relaxed, stop sneezing, coughing, and their blood pressure gets normal. In a few days they are ready for their surgeries." The first thing I thought, was: "It's probably because there are so many needles that eventually some of them reach proper acupuncture points." Later I discovered that the inventor's logic had indeed worked exactly this way.

Kuznetsov's story

Here is how the story transpired in the late 1970s. The inventor, Ivan Kuznetsov, was a music teacher in a kindergarten. He was the only man on the kindergarten's teaching team that was why one day he was asked to take care of insect treatment of the building. They supplied him with a gas mask and a coverall, but had forgotten to warn him about wearing gloves. Because he was not a professional exterminator, he did all this toxic work with bare hands, and was severely poisoned. He had chronic pain; his impaired peripheral circulation often resulted in muscle spasms so severe that he was unable to move his limbs for weeks. His kidneys practically failed.

The only measure that worked for him was acupuncture, but this Eastern modality was just making its way to Russia and only a few outpatient clinics offered the treatment for free. As to private practitioners, they charged fortunes and he could not afford as many sessions as was required. For this reason he decided to learn how to perform acupuncture on himself. In this endeavor he failed, because, he said, many points he needed were on unreachable parts of his back. Instead, he invented his Applicator, piercing a sheet of tire rubber with office pins, 1/4" apart, to lay on with his back.

The original version of IPLIKATOR with metal needles

Luckily, the Law of Physics did its job well and with about a thousand single needles sharing the body weight none pierced the skin.

As crazy as idea seemed -- it worked!
Ivan Kuznetsov filed his invention claim in 1979 and had it granted in 1980. In 1980, a publication about this invention was released in the very central newspaper of the former USSR - "Pravda" (The Truth).

By the way, there were actually two central official (unofficial just didn't exist) newspapers: the Truth and the News. Everybody knew a caustic joke about them: "The News is no truth and the Truth is no news." People couldn't help saying this joke sometimes even jeopardizing their safety: one could get the KGB's attention for lesser a reason.

Thousands of tests in clinics
In 1981 the invention was described in the popular science magazine "Invention and Rationalization." After that, a number of research institutes in Moscow conducted unofficial clinical trials, with results that have never been published. Here I need to comment on the mystery American term "unofficial clinical trial." There was no such thing as the FDA in the former USSR, there was a Department of Public Health and the minister of this department allowed the device to be manufactured and sold with no trials whatsoever. So the doctors interested in the method were free to try it on their patients whether they wanted it or not (anyway, nobody ever complained). There were quite a few professors in Moscow research clinics that did try it: In May of 1981, the State Institute of Physical Culture conducted 2450 sessions with the applicator. In 1982 the Institute of Neurosurgery recorded the results of 1000 sessions on 30 patients and the Institute of Experimental Surgery tested 75 patients in 750 sessions. In 1983, the Central Institute of Trauma tried the device on 176 patients and the Department of Facultative Surgery of the Second Moscow Medical School investigated the method's effects on the electrocardiograms of 120 patients with heart diseases.

Medical Newspaper
In 1987 the "Medical Newspaper" published an article describing the amazing case the reporter witnessed in Kuznetsov's office: a lady who had suffered from insomnia for years fell asleep after 15 minutes on the Applicator. The author interviewed several medical doctors asking them how they thought the device worked. The answers were quite general ones like "The method aims to increase unspecific stimulation flow to the brain areas controlling

general alertness. " Sample articles from The Health Magazine can be found in the APPENDIX.

Insomnia, colds and flu, asthma, and panic attacks.
About that time, I also started using the device for my insomnia (it invariably worked) and after several weeks I realized that my previously frequent colds and flu subsided. I was also able to help a good friend of mine with severe asthma who had not been able to sleep in his bed for several months and had to spend nights sitting in an armchair. He was scheduled to visit a doctor for a second opinion and was asked to abstain from his usual asthma medication (broncho-dilators). As a result, he could not withstand the small physical effort needed to get on his feet to walk to the car. I made him lean forward, spread the Applicator upon his bare back skin, pressed hard and held the pressure for about 20 minutes. Gradually, I could hear his wheezing disappearing, his skin color improved, and he stood up cautiously, made a step forward and smiled: "I hope this medication has not been prohibited!"

The Applicator has also worked for my panic attacks. Now, I am going to tell you about an unexpected and incredible consequence of using the Applicator for this particular malady. Believe me I would never risk my reputation telling this story unless it had been well documented by doctors.

During the 1991 coup in Moscow
It happened during the 1991 coup in Moscow. I don't need to tell you how much stress there was for all of us and how much worse my panic attacks had gotten because of the situation. So, I used the Applicator several times a day. At the time, I caught flu or what I thought was flu: I had a bad fever, cough, etc. In a week the flu

symptoms disappeared. At this point I need to tell you that my family was preparing to leave for the US as permanent residents (we had obtained our green card because of my husband employment offer by Duke Medical Center). As a result, we had to undergo a medical exam including TB tests and chest x-rays that had to be evaluated by American doctors. We all were given a clean bill of health. I will refer now back to my flu; I had found it very helpful for my fever and cough to use the Applicator. It was in September 1991, and in February 1992, already in North Carolina, I had to repeat my TB skin test as a hospice volunteer. The result was a disaster. I thought I would lose my forearm it became so inflamed and swollen. I rushed to the Duke medical center; my chest pictures were taken again, after which two doctors came in the exam room reproaching me for taking the TB skin test. "After having tuberculosis, you should not ever have the skin tests," they said.

I was shocked: "What tuberculosis?! All my tests were squeaky clean in August!" "Well," they said, "Apparently you had it later. What we could see on your pictures today is a well cured tiny spot in your lung telling us the treatment was timely and effective." I failed to convince them I had no treatment whatsoever but the Applicator. They strongly advised me on regular chest exams and against TB skin tests. Later, I had to go for my chest exams twice a year because I worked in a rehab center and then with newborn premature babies, and all my pictures remained clean till doctors stopped noticing the "well cured spot" in my lung at all.

In my early experiences with the Applicator, I figured out that it worked for most everyone's headaches. It worked for my son's respiratory allergy, which was so severe that he had to be home schooled for 2.5 years. One of my friends suffered from some kind of undiagnosed gastrointestinal discomfort for years. After 2 month of laying regularly with her stomach down on the applicator, not only did she finally find pain relief, but also, she could wear her bikini again because of the noticeable toning of her skin.

My husband inherited bad varicose veins complicated by weight lifting when he was in his 20s. After a severe accident with trauma and several surgeries, his condition deteriorated: veins swelled, inflamed, he had muscle cramps and was in extreme pain. He was told that he had no choice but surgery. It took us 1.5 months of intensive Applicator treatment to heal his veins, though not completely, to the point he did require surgery anymore. His pain, cramps, and swelling disappeared. Now, when they are back, we know what to do. Now it takes him half an hour to stop pain and cramps any time they occur.

Kuznetsov's Team
In the late 1980s, I was working on a database on alternative, folk and other unconventional methods of healing. That was when I first met Kuznetsov's team. The core group was made up of people with very different backgrounds many far from medicine at all, there were also a few medical doctors there, working part time. I'd been introduced to the professor of endocrinology Vladimir Snegirev who came one day to the office to protest against the method because his diabetic patients were quitting insulin. However, after Dr. Snegirev investigated a number of cases, he was so convinced that he joined Kuznetsov's team, too. "What the

Kuznetsov's people are saying about the method is awful," complained he to me, "The medical illiteracy of theirs is not even funny. But when you look at the results, at all the hundreds who are getting so much relief, all you can do is set aside your medical literacy and work hard to promote the method." What he meant by illiteracy was their statement the Applicator cures all, but we both heard a few of the group's other favorite clauses, or rather incantations, like:

"All the filth of your disease has to drip out of your body through the needles," "The needles connect you with energy channels of the Universe,"

"The needles are antennas to receive the healing commands from the Cosmos," etc. Or was it an intuitive approach to imply Guided Imagery and Faith Healing?

In time I met Mr. Kuznetsov personally. In his 80s, he looked incredibly robust and merrily aggressive, you could never tell that two decades go he was practically handicapped. Most of all, I had dealings with Valeri Poukov, the engineer in the team (after I.I.Kuznetsov's death in 2005, the method continued developing under Valeri Poukov's leadership (more information at regeneration.ru)

I also met a neurologist, a physician, and nurses who worked for the team and everybody told me approximately the same story of disbelief and conviction. They let me look through some files consisting mostly of people's letters written to Ivan Kuznetsov containing success stories. Nobody among the stuff bothered analyzing the invaluable information concealed dormant in these files.

**Arthritis, Leukemia, or Cerebral Palsy.
What Do They Have in Common?**
Medical conditions I discovered while looking through the letters:

1. Arthritis

2. Bone fractures

3. Cerebral palsy

4. Frequent colds and flu

5. High blood pressure

6. Low blood pressure

7. Leukemia

8. Osteomielitis

9. Insomnia

10. Asthma

11. Diabetes

12. Breast cancer

13. Gastric ulcers

14. Sarcoma

15. Radiation syndrome

16. Radiation therapy syndrome

17. Chemotherapy syndrome

18. Terminal cancer (!)

"Very simple: it's a panacea"
To my question about the working mechanism of the applicator, Valeri Poukov replied "Very simple: it's a panacea." That idea was very good, indeed. Really -- what do diseases such as diabetes, sarcoma, osteomyelitis, radiation syndrome, infertility and frequent colds have in common?

For example, there was Lena's case. She was a fourteen-year-old girl, who had been born with cerebral palsy. She had never learned how to walk. When her mother learned of the method, she bought several of the devices, in order for Lena's whole body to benefit from the treatment. Lena worked hard lying on them several hours a day and, after a year and a half, it happened! Lena was able to stand up and keep her balance. Then she started walking. I saw Lena take her first steps on videotape, shaky and awkward. Lena's mother with tears of happiness in her eyes was ready to encourage and protect her daughter at that moment as in the past. I met Lena's mother in Kuznetsov's office. She began crying as soon as we started talking about Lena's success. "Just imagine, what we could have achieved if we had known about the method when Lena was a baby!"

Chernobyl's babies
Speaking of babies, I have read something even more impressive, the claims being made relative to babies that were born to survivors of the Chernobyl disaster who used the device. Valeri Poukov told me that newborns of Chernobyl's victims, if wrapped from the very birth into tiny special vests with tiny plastic needles inside, fixed tight on the body to avoid scratching, fared significantly better than newborns from the same population who were not treated. What's more, the rate of birth defects was much lower among Chernobyl mothers who used the device during pregnancy.

Radiation therapy
These results seem to be consistent with Poukov's data on using the method in conjunction with radiation therapy. Patients who used

the device had much less discomfort, had less hair loss, and had less irritated intestines and skin. Inspired by these effects, some of them continued using the device between therapy sessions. In the long run, these patients showed faster recuperation and less recurrent tumors.

Sarcoma
Another case involved a teenage girl with sarcoma of the shoulder joint. I saw a set of x-rays taken at six-month intervals. On the earlier one, the shoulder joint with the sarcoma was a complete mess. The doctors' prognosis was far from optimistic and she wanted to try using Kuznetsov's device. Although they could not promise her more than a few months to live, the medical staff prohibited her from using the device. But each night she secretly slept on the "bed of nails." After six months, even an inexperienced radiologist like me could see the improvement on the x-rays. Seven months later, the girl was playing her guitar! I'd been given the chance to view the before and after x-rays. I am not a specialist in x-ray pictures, but the absolutely remarkable improvement was evident even to me.

Hypotheses Needed
When I asked about the scientific principles underlying these results ("Please other then panacea stuff!"), Valeri handed me a folder with copies of the conclusions of investigations conducted by five Moscow clinical institutes. By the way, I had to write them down by hand in his office, because Valeri couldn't allow me to make copies: at this time, to make copies in Russia, one needed to obtain a special permit from the censoring department of the police. These summaries are the only medical documents I can

refer to. Let us put it as an appendix for those people who wouldn't be bored by reading quite specific medical information.

By the time I had seen all this information, I was not too surprised to find my ten year study had not uncovered many failures -- no matter what ailments were dealt with using the applicator: from insomnia to asthma!

In the early 1990s, I was employed at the University of North Carolina -Chapel Hill in the physical therapy department and for the first time I encountered the necessity for some kind of "scientific" explanation when I was asked to give a presentation on the Applicator. You see, being an enthusiast, I showed the Applicator off, talked about it, cured head- and backaches with it. But the scientific explanation (finally!) needed a scientific approach.

I went to the library and began the search. That was years ago now. Since then I started my own private practice as a health consultant, and found Americans who believed in the method. We tried to order some Applicators from Russia but failed, so we had no choice but to develop and start producing the device in the US (we named it "Panacea" in Valeri Poukov's honor) but this is another story.

Now I feel I can put forward some hypotheses I consider well developed. There are basically two, and one of them is not exactly mine, but both are related to problems I came in contact with back in Moscow.

Self-acupuncture?

First of all, the prototype was made out of vacuum resin with a large number of office pins in it. The hope was to find the acupuncture points with some of the pins, like shooting a penny with a large number of pellets. The ideal was that the many pins not on target would not harm whatever good the right pins might do. It is possible to say that these hopes came true, since Mr. Kuznetsov healed himself then thousands of other suffering people who lived to thank him. Now, one can call the method "Self-acupuncture."

Can acupoints explain the effects?

If you take a look at any acupuncture meridian map, you will see that in the most usual supine position with the Applicator under one's back, the acupoints involved can be expected, according to a Chinese atlas, to improve the following body functions:

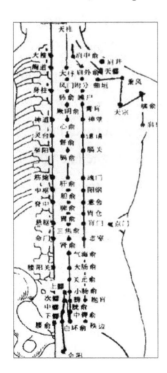

- Strengthening the liver, spleen and kidney.
- Alleviating headaches, fatigue
- depression and insomnia.
- Activating the immune system.
- Relieving flu, cold and asthma.
- Regulating digestion and elimination.
- Improving the conditions of cystitis, diarrhea, hemorrhoids, PMS, and complicated periods.
- Easing spinal problems, sciatica, muscle spasms and cramps.

This is all wonderful, but the question still exists on just how acupuncture works. Aside from metaphysical explanations about cosmic energy flow through the channel meridians, and despite the well-proven fact that acupuncture really works, there are very few proven theories.

There is a principle: not to pile on more theories than necessary. In other words, while there are still simple answers (in this case, physiological) not to bring in such exotic theories as cosmic energy. To follow that principle means to stay within the scientific paradigm.

Self-diagnosis, Neurophysiology, and Neuro-computers.
On my opinion, the best of the few known theories on acupuncture belongs to the team of theoretical biologists under the command of the professor of the Russian Academy of Sciences, Dr. Dmitri Chernavski. Neurophysiologist Dr. Rodschtat and Victoria Carp, a specialist in computer-recognition of objects, are both on the team. I have known Dr. Chernavski for many years, this great person (a personal friend of the late Andrew Sakharov) and outstanding scientist well deserves to be the hero of a novel. In the late 80's, I was working on database on neural networks for his team's project on "neuro-computing." What was it about?

Formally, simplified models of neural structures in a form of electronic devices existed for a long time, starting a whole new class of computers with artificial intelligence - ones that can learn, recognize objects, and correct their own mistakes. Sounds like a paradox, but the opposite idea to explain the mechanisms of a live brain using the known electronic models proved to be useful. Thus the theory of the self-diagnosing function of an organism was developed. Most of these self-diagnose recognition processes take place in the spinal cord.

Skin, spinal cord, and sympathetic nervous system

Information from the skin enters the spinal cord through, is being processed there and exit the spinal cord to join the sympathetic nervous system and reach the internal organs.

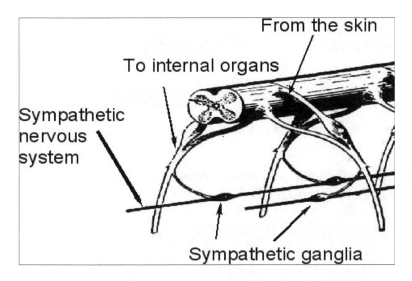

In the gray matter of the spinal cord (see picture on the page 18), the neurons are organized into conglomerates, the so-called "Rexed Laminae"- numbers I to X on the picture:

Their functions are well known. The signals, separate from each other, from the internal organs, as well as from the skin and muscles, initially go through the lamina I. Then the

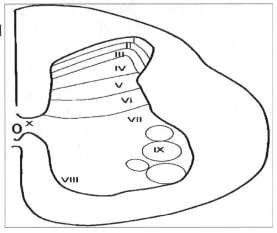

signal moves through the lamina II, III, and so on, while increasingly interacting with each other and, after laminae X, finally reaching the brain in the form of an integrated piece of information about the body's state of being. The computers that recognize objects have basically the same structure and signal integration. In cases, an omitted signal or one that is not strong enough will be compensated for by another one, thus fixing the mistake.

According to Dr. Chernavski, skin stimulation at the point of acupuncture accomplishes the same goal - it increases the flow of signals from an organ adding to this flow signals from the corresponding point on the skin.

The skin projections of internal organs (zones of Zakharian-Head)

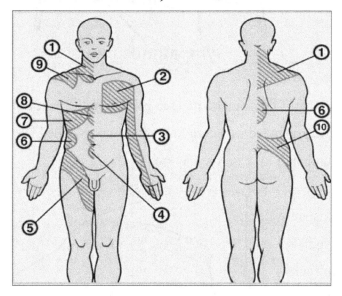

1) heart; 2) intestines; 3) bladder; 4) urethra; 5) kidneys; **6) liver**; 7) stomach; 8) uro-genital system; 10) pancreas.

But what does make the signals from acupoints target the same spinal cord centers as signals from an organ do? I like the theory of embryo-genesis. It states that two sets of body tissues - the organ's and the corresponding skin point's - once originated from similar maternal cell groups (which has been completely and directly proven). Later, being separated in the course of organogenesis, they nevertheless kept their primordial memories (which is true in the sense of their common projections into the brain centers). These centers may not be "awake" enough to pay attention to a weak signal from an organ but they can be "awaked" by additional stimulation, from the corresponding skin point projecting into this particular center

It is as if you want to send a letter to someone down a stream with little or no water. You fold this letter into a paper boat and face the fact that there is not enough water to carry it. Add some water, and the paper boat will get there. The method of adding the water, the amount of water, or the paper of which the boat is made – all these have no effect on the content of the letter you're sending. The same way, the "illness letter" from a diseased organ to the brain can reach the diagnostic center in the brain only if it is carried within sufficient amount of signals in the "stream" of bodily information. Acupuncture or for that matter the Applicator can contribute significantly to this common "stream" and thus help in carrying the "illness letter."

What happens when the disease is recognized, the theory of self-diagnosis does not explain that the body has enough resources to battle the disease on its own. Conventional medicine rejects that statement, while holistic medicine thrives on it. I think conventional doctors make the mistake of disregarding invaluable information including that collected within the strict Western science paradigm. Lets' take a look at, say, skin stimulation at large. Why do you think children rub their bruises?

Why Do Children Rub Their Bruises?
In many religions people practice intense skin stimulation to reach a heightened state of spirit. They wear rude-fabric clothes and prickly necklaces, they pray kneeling on dry pees or grains, and they lash themselves with whips. Yogis walk upon burning charcoals and lie down on a bed of nails. Julius Caesar would order his skin pinched to relieve his neuralgia. An ancient Roman physician Pliny used skin stimulation for treating asthmatic patients. Ancients Greeks elaborated special rules for rubbing their patients' and athletes' skin for curing diseases and improving performance.

American physician Dr. Zalmanov used skin stimulation with hot water; he called capillaro-therapy, for a wide range of diseases including those resistant to any conventional treatments. Swedish doctor Paavo Airola advises dry brush massage of the entire body for many ailments and as a general disease-preventive measure. A Russian folk healer and philosopher, Porphyry Ivanov, cured practically everything, including mental retardation, with skin stimulation by cold water.

The skin-health relationship may even consist of more than that. Different skin areas are not equal in this respect. The Chinese have known this for more than 4,500 or even 5,000 years. The earliest written document on the tradition of acupuncture is believed to have belonged to the Emperor Huang Ti. The ancient Chinese learned how to diagnose illnesses and how to treat the internal organs and body functions through their matching points on the skin. These special areas and points of increased sensitivity and reactivity on the skin can be affected by a number of techniques known as acupuncture, Jin Shin Do, Shiatsu, zone therapy, reflexology etc. In Russia, all these techniques have a common name: "Reflexo-therapy", from the word reflex. This word means: you make an action and receive a certain response, like sending the ray of a flashlight onto a bathtub water surface and watching the spotlight reflected on to the ceiling. This is basically how it happens in all these techniques: a healer affects a skin point and triggers a reflective physiological reaction affecting an internal organ.

The Largest Organ of Your Body May Be Starving.

At one time in our evolutionary history the skin played a much more important role in the self-healing capability of our bodies than it does today. Over the last thousands years or so (nothing on the evolution's scales), the skin has been experiencing "informational starvation" thanks to the use of the clothing we wear and the controlled environments in which we live and work. The skin, the largest organ of our body, is not having as much stimulation as it is designed to have. Stimulation of the skin causes the secretion of many regulatory substances, which control physiologic functions.

Well known are endorphins - chemicals released by our bodies occurring naturally as a response to any deviation from the "normal" state of the body. Among the many triggers of such a release are pain, sex, novelty, blood loss, and lack of oxygen, stress, fasting, exercise, and delicious foods. Recently, it has been stated that acupuncture and actually any intense skin stimulation will cause a significant release of endorphins. The pain killing effect of these are without a doubt based on that phenomenon.

One of my projects as a researcher was a review on a book about regulatory peptides including endorphins. In the book, my co-authors and I described a curious state the body falls into after the endorphin concentration has gone up: a number of other physiological regulators are released into the bloodstream, among which are such well known ones as growth hormone and insulin. Each one of those regulators changes a whole number of the body's functions. As a result, it is not surprising that very serious diseases are linked to abnormalities in the endorphin system: addiction, schizophrenia, epilepsy and Parkinson's disease along with many others not least of all PMS and weight problems.

Body awareness
Thus, skin stimulation, even not necessarily as accurate as in acupuncture, but intensive enough, does two things:

1. Points out the inner needs of the body to itself. That is important because today's civilized man has lost most of his "body awareness"
2. Ensures the rearranging of body's functions under the control of endorphins. The "aware" body will do its own work to heal itself.

Research: the Pilot Study
In 1995, I came to work with the Community Holistic Health Center at Hillsbourough NC (currently, the Center moved to Hillsborough.)

At that time, the Center's board of directors was about making its strategic development plans and I suggested a research program on Reflexo-therapy. After consulting a lawyer and getting information on human subjects rules from the University of North Carolina, Chapel Hill, and after practically all board members and stuff tried and loved the applicator, the project was accepted and in 1997 we had our pilot study completed. We followed up 200 of my patients who used the Applicator and asked them to fill out our simple questionnaire. The amazingly many of them, 126, answered our questions.

The Panacea mat used in our pilot study in 1995

The 20" x 9" Panacea mat is made of flexible polyurethane and contains 16 pointed sharp "stimulators" per every square inch. The flexibility allows to evenly press the mat to body's curved surfaces.

A closer look at the Panacea mat

Results

The five most frequently reported reasons for using the Applicator were stress, pain, muscle spasms, mood swings and insomnia. Overall, we had an astonishing spectrum of reasons why people were using the Applicator: lack of energy, mood swings, insomnia, impotence, asthma, varicose veins, hypoglycemia, skin condition, inflammation, weight problem, fast heart beat, nasal or sinus congestion, fear or panic attack, allergies:

1. stress= 47 cases
2. pain =40 cases
3. muscle spasms=20 cases
4. mood swings, irritability=13 cases
5. weight loss=12 cases
6. nasal or sinus congestion=10 cases
7. fear, panic attack =10 cases
8. insomnia=10 cases
9. fast heart beat=9 cases
10. skin condition=8 cases
11. inflammation=6 cases
12. impotence =8 cases
13. varicose veins=3 cases
14. hypoglycemia=2 cases
15. tremor (Parkinson)=2 cases

Out of our 126 subjects, 98% reported pain relief, 96% reported relaxation, 94% improvement in the quality of sleep, and 81% reported an increase in energy level.

Approximately half of the subjects with allergy problems reported their symptoms' relief. Among those who tested the method while having no particular health problems, more than a half nevertheless reported one or more positive effects of the Applicator.

Effects of applicator

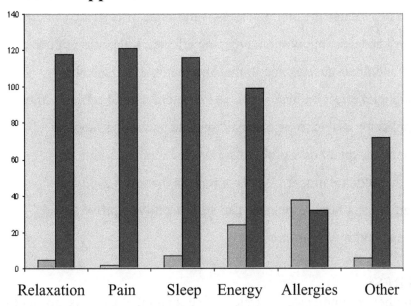

Black bars – improvement; gray bars – no effect

Major Practical Application

PAIN
All types of pain are reduced or eliminated. Relief of headaches and back pain is the most reported benefits. We tell people that a headache is not a disease but rather a symptom of a number of different ailments. Most frequently these are: muscle tension, abnormal vascular reactions, stress, allergies, blood clotting and neuralgia. Migraine headaches have a biochemical cause, for example, excessive secretion of Serotonin and Prostaglandin by the brain and blood vessels. The most common reasons are muscle tension and spinal disk misalignment.

Common painkillers may only relieve the acute pain sensation. None of them removes the cause of pain. Some painkillers containing caffeine may help to dilate contracted blood vessels, but they may also cause unpleasant side effects on your heart and blood pressure. The Applicator increases blood flow to a target area and dilates the blood vessels -- the same as caffeine can do for your headache, but without any side effects. Relaxation of neck and shoulder muscles due to the Applicator action also has a beneficial effect on headaches. As a potent natural relaxant, the Applicator is able to ease spastic muscle backaches, and for managing spinal disk problems, it is an excellent addition to chiropractic treatment. We have reports that even 2-3 weeks' regular use of the Applicator can significantly improve spinal adjustment by a chiropractor.

STRESS
Another common case is stress. We explain that stress is a psycho-physiological reaction of the human organism to above-normal effort. Stress is not a disease: a fit and well-balanced body is able to withstand health conditions resulting from stress. Any efforts, which exceed a person's level of tolerance, can bring on stress. Exercise is stressful for the sedentary, but is easily tolerated by physically trained people. Likewise, socializing may be very stressful for shy person, but fun for an extrovert. Among the severe results of sustained stress are cardiovascular diseases and immune system depression.

Stress itself is a potent releaser of endorphins. However, when the body is out of balance due to sustained stress, it often experiences an endorphin deficit. Then it needs help to fight stress. The endorphins released during the Applicator session have been shown to calm our emergency nervous system, the Sympathetic one, and balance it with its antagonist, the Parasympathetic Nervous System. This balance is one of the most crucial conditions for your general health.

EXERCISE
By helping endorphin release, the Applicator makes our bodies happier about physical efforts. The Applicator prepares the body for a workout by enhancing metabolism. Our research shows significantly heightened metabolism due to the Applicator treatment. We have information on increased oxygen consumption and additional energy expenditure due to increased body-heat dissipation through the skin.

Since the Growth Hormone (GH) is so popular we did some research on possible involvement of the Applicator with it. The human body produces GH throughout its lifetime, however, secretion decreases significantly with age. GH deficiency causes inability to perform physically, plus contributes to cardiovascular diseases, osteoporosis, etc. GH replacement therapy has become a very fashionable (and expensive) tool in the struggle against aging. Many advantages of this approach are clinically proven. Due to its prominent anabolic properties, GH stimulates lipolysis and protein synthesis, resulting in decreased body fat and increased muscle mass.

GH provides an overall improvement of the immune system and mental abilities and also enhances collagen synthesis that leads to improved skin condition. It has recently been concluded that, instead of direct GH injections, the use of natural GH releasers can be beneficial to the human body. Exercising, fasting, and vegetarian dieting are among the known natural GH releasers. Very strenuous exercise has been shown to increase the concentration up to 2,000 times -- naturally! Furthermore, a number of substances naturally occur in our body that can cause the release of GH through biochemical pathways. Some amino acids, for example, Arginin and Ornitin, are used as standard diagnostic tools in clinics to test the body's ability to release GH. Release of GH can be elicited by endorphins, too.

The Applicator treatment significantly increases blood flow to an overworked muscle. Applicator use improves metabolic conditions, oxygenation, toxin evacuation and delivery of nutrients, thus, eliminating the cause of muscle spasms. We have numerous clinical and personal reports on fast spasm relief with the Applicator.

WEIGHT LOSS
Another application that we think of is weight management. Here is why. We overeat to comfort ourselves during periods of stress, pain, or boredom. Tasty foods release endorphins into the blood stream. The Applicator can cause release of these same endorphins without the harmful side effects of over-eating or increased weight.

We maintain a wrong weight, because our set point for body-weight is elevated above normal. The set point is a special arrangement in the brain (hypothalamus) nuclei, which controls our appetite. The neurons in these centers are sensitive to many biochemical constituents, including endorphins. The Applicator can supply additional endorphins to adjust the set point. They always work to make the correct adjustment. For example, clinical data show that both conditions of obesity and anorexia correspond to endorphin imbalance in the blood. By normalizing the body's set point, the Applicator may help to avoid yo-yo weight changes, which is the usual problem after dieting.

The Applicator helps optimize health by triggering a cascade of important regulators. Skin stimulation causes endorphins release, and initiates secretion of many regulatory substances, such as the potent appetite suppresser (neuropeptide Cholecystokinin), and an important enhancer of metabolism thyroid stimulating hormone, to name just two. GH is yet another secreted regulator, and its antagonist peptide Somatostatin is suppressed by the endorphins, leading to even further GH enhancement.

MEN'S HEALTH:
An Interesting Case of Cured Impotence

"I used this gadget for my neck and shoulder pain after my car accident. It helped right away. After two months of use, I forgot about my insomnia and -- beyond any hopes -- my impotence of 5 years was gone!" -- Michael was the first person who shared with us the Reflexo-therapy (Panacea device) effect on impotence. Now we have 8 cases of reversed impotence. Michael was the first person who shared with us the Reflexo-therapy effect on impotence. Now we have 8 cases of reversed impotence.

WOMEN'S HEALTH (PMS and hot flashes)

Hot flashes are among the symptoms of menopause. In the US, about 85% of women experience hot flashes and 40% of them consider this ailment severe enough to commit to estrogen replacement therapy. However, Asian women are luckier. This fact is usually explained by the difference in traditional diets, which are higher in soy products in Oriental cuisine. Thus, it is currently advised to include tofu, miso, soy milk, etc., in middle age women's everyday ration. There is also data on favorable effects of plant analogs of estrogen found in herbs like Licorice, Wild Yams, Angelica, Blue Cohosh, etc.

Clinical research has shown that during a hot flash attack the concentration of endorphins in the blood is significantly decreased. This is why we offered the method of self-help Reflexo-therapy to several hot flashes sufferers and got very promising results. There is direct scientific evidence that endorphins can control hot flashes through the inhibition of the female hormone Luteotropin. Heart palpitations, anxiety, and a sensation of heat usually accompany hot flashes. Endorphins calm the sympathetic nervous system, eliminating therefore abnormal heart beat and other vegetative expressions of nervous tension. Endorphins are clinically proven to be effective in body temperature regulation: they increase body temperature if it is below the normal level, and decrease it during fever.

Among numerous symptoms of PMS are the following: anxiety, irritability, mood swings, fatigue, cravings, headaches, water retention, depression, memory instability, and various skin problems. Cramps during PMS are the result of uterine muscle constriction that causes diminished circulation, insufficient oxygenation, and increased lactic acid concentration. Clinical experiments show that PMS corresponds to endorphin disturbance. On the other hand, there numerous experimental data have shown endorphins to be powerful tissue protectors against hypoxia (lack of oxygen).

We can be very positive that intensive skin stimulation helps to improve local circulation and restore oxygenation. The Applicator definitely provides this type of stimulation. Considering all this, we asked several PMS sufferers to try the Reflexo-therapy method in hope that skin stimulation would release endorphins into the blood and normalize their concentration. The results were encouraging. It helped to relieve pain and cramps as well as mood swings and headache.

Case histories.
Carolyn S., Hillsborough, NC 51 years old professional tennis player. Reason for using the Applicator: muscle, back and joint pain due to professional repetitive trauma, failing to get help from chiropractic, all kind of massage therapies, yoga, and acupuncture
Tolerability: she had no problem with the device from the very first time of using, as well as her 9 and 14 years old children.
Reaction after first use: loosening of the muscles, sleepiness
Reaction after three weeks of use: alleviating the spinal and spasm problems, stress reduction
Comments:
Carolyn describes her relations with the Applicator as a benign addiction. She became the method's proponent and succeeded to convince her husband, an MD professor, to try it and later to recommend it to his patients.

Rachel R. (Raleigh, NC) Biofeedback practitioner described her experience with training a patient while using the Applicator.

Reason for using the Applicator: in order to enhance the effect of biofeedback, as a device aimed to improve body awareness and increase local blood flow.

Tolerability: Good -- the patient was concentrated on the biofeedback task and followed verbal instruction to relax and accept the device. Reaction after the first use: prior to implementing the Applicator, the patient was not able to control finger temperature and she managed the task immediately after.

Comments: 1) Rachel commented that the skill that usually requires a week session of training or longer occurred instantly during the Applicator treatment. 2) The patient explained that she could not understand first where to take the heat from to elevate her finger temperature, and immediately after laying down on the Applicator she felt the device as a source of energy and heat, so all she had to do was to relax and allow the heat to reach her finger.

Yadviga D. (Chapel Hill, NC) 41 years old.
Reason for using the Applicator: thyroid problem, tachycardia, morning sickness and lack of energy
Tolerability: she had no problem from the very first time
Reaction after the first use: heart rate slowed down after 30 min. on the Applicator. Reaction after three weeks of use: energy increase in the morning, slowing the heart rate every time when tachycardia occurred
Comments:
Yadviga was first who noticed energy increase after 5 to 10 min. versus 20 to 30 min. when relaxation began. Later many sportsmen reported this two-wave character of the effect. We investigated literature data on the time course of endorphins'

physiological action and have found our results to be in accordance with this time course.

Michael U., New York, NY. 72 years old

Reason for using the Applicator: back and shoulder pain after car accident two years ago.

Tolerability: during the first several sessions with the Applicator, he had to use it through a tee-shirt, then through a silk scarf, but he felt that the method started working for him only after he used it on bare skin.

Reaction after first use: general relaxation and drowsiness

Reaction after three weeks of use: reduced intake of pain and insomnia medicine, reduced night incontinence, reversed impotence

Comments: 1) Michael discussed the use of the method with his neurologist, urologist, orthopedist, and physical therapist. None of them advised him against using the Applicator, however, none of them expressed any interest to the method. 2) Michael was the first person who shared with us the Applicator's effect on impotence. Now we have 8 cases of reversed impotence.

Reasons for using Panacea

- Osteochondrosis (Paula, 70): after first treatments with the Applicator reported that pain relief was more significant than with drugs, her sleep has been improved.
- Parkinson disease (Tatyana, 74): after 6 weeks side effects of levo-Dopa corresponded to the drug's time-course -- 30 minutes before and after the next pill -- has been diminished. Shakiness, stiffness and anxiety were all relieved.
- Hypoglycemia (Susan, 42): reported eliminated cravings and shakiness occurred due to lowering blood sugar level.

- Allergy (Constance, 47): reported significant nasal decongestion effect during the first and all consequent Applicator sessions.
- Spinal Problems (Bill, 51; Steve, 33; Jim, 50s, Pam 20s, Ludmila, 60, Elfriede, 50+, Ingrid, 60, Helen, 55, and many others): as a short term response, reported significant pain and stiffness reduction. Long term effect in some cases was facilitated response to chiropractor's adjustment.
- Sports & Fitness (Sonny, 57, professional senior body-builder): body fat decreased from 18.6% to 7.5% during 6 months of the treatment. Muscle mass has been increased. Subjectively he reported decreased cravings.
- Muscle pain and cramps (Dance Company "Constellation", 15 girls in their 20s): reported completely disappeared muscle cramps and joint pains.
- Inflammatory Pelvic Syndrome (Victoria, 35): in three days she reported completely disappeared inflammation, fever and pain, doctor's visit after 2 more days of treatment revealed no signs of the disease.
- Skin (Stephanie, 7, a patient of Dr. Dennis Fera, a physician at the Community Holistic Health Center, Carrboro, North Carolina): skin scar tissue discoloration decreased, elasticity improved; (Tatyana, 47): complexion improved; (Jack, 60s, Vietnam veteran, suffered from severe skin problems due to the Orange gas poisoning): After 2 months of using the Applicator had visible improvement in the skin complexion. (Elena, 55): used the Applicator for her stomach pain and besides the pain relief, reported significant improvement in the skin tone.
- Night sweats (Hannibal, 70): had a heart attack after which felt permanent weakness accompanied with sweating especially

profound at nights. After the first use of the Applicator and until now never experienced the night sweating.

- <u>Fibromyalgia</u> (Sandy, in her mid-thirties, has been diagnosed a year ago): Especially suffered from muscle pain at night preventing her from restful sleep. Slept well during the very first night staying on the Applicator for several hours.
- <u>Depression</u> (secretary of Dr. Stopford, Duke University, 40+): The degree of depression has arisen the question of her disability. Applicator started working for her gradually, eventually allowing her to concentrate and to work.
- <u>Asthma</u> (Andrew, 60s, Michael 13): significant relief of symptoms after the first use, Andrew has been able to sleep in the bed instead of an armchair, Michael was free of the disease after 6 months of use.

Testimonials of Users of the Panacea Concerning Its Effectiveness

76 year old couple: "We both use it regularly together with our walking, exercise and diet along with vitamins and minerals. The Panacea] relaxes and stimulates endorphins. We are completely sold on the 'torture pad'!"

Lawyer, tennis player and ex-competitive swimmer: "[By the fifth time I used the Panacea], the period I felt uncomfortable had shortened dramatically. For those with stiff morning muscles, it's great. Its great for insomnia, too -- diverts the mind and relaxes the body. I was one of your doubting Thomases. But the Panacea is a terrifically effective, efficient muscle relaxer. It has banished my morning back stiffness and iron outs pulled back muscles. I have used it twice when I had trouble sleeping and it truly relaxed me to the edge of sleep How strange something so simple works so well."

Professional massage therapist: " I use the Panacea in the treatment of my clients. It makes my work as a massage therapist significantly easier."

Tennis pro: " I use it if I'm feeling stressed out or in pain. It is uncomfortable at first, but within 5 minutes I start to feel mellow and very relaxed."

Dance instructor: "There is a real feeling of "relaxed" muscles – it redistributed the tension in my lower back. It1s a clear source to release specific muscle tension."

Singer: "I try not to think of the brief discomfort: I focus in something else. I have really noticed a dramatic difference in relation to all severe headaches I was having. They seem to just fade away the longer I use the product. Thanks a million!!"

Seamstress: "I spent 40 minutes on the Panacea each night before sleep. It does wonders for a headache."

(ET): "I'm hooked -- this is a great addition to my wellness regime. I wish it covered a larger portion of my body."

Sales manager: "It takes a while to adjust to the Panacea. Also, placement on the back is important. Sometimes I have to adjust several times before I can get a comfort level to withstand. Breathing is also very important. It helps to have the tape on where they tell you to breathe slowly and deep. This adds to the ability to relax into Panacea."

Tennis instructor: "I sometimes need to shift myself for points to hit the correct spots. [Immediately after using the Panacea] I feel more focused and relaxed."

Psychologist: " I have used the Panacea on my thumbs for tendonitis and for headaches including sinus pain. Fabulous results! The combination of massage therapy and the {panacea has been the only thing that has actually helped my lower back. My riding (horses) is once again possible and better than it's been since I was a teenager (and I just turned 50!)

Manager: " Along with physical stress, I suffer from severe nightmares, which inhibits my ability to go to sleep greatly. The first night I tried the mat, I stayed on it for about twenty minutes. My skin was fire-hot for about ten minutes afterwards, then felt wonderfully relaxed. I slept through the night for the first time (without medication) in almost a year.

Dancer: " I'm sleeping longer and more deeply and have had healing of an arthritic hip joint. I have less tension, pain, and fatigue during the day.

Car sales person: "I always feel very relaxed after using it. Now I have to use it because if I don1t my body doesn1t feel right.

Writer: "The sensation of "first pain" grew less over time. At first I was uncomfortable for 3-4 minutes. Then (after a week) it took only one minute. I really love the warm rush after that.

Dancer: "I use for a severe hamstring injury and insomnia. It is great for pain, injuries, headaches, energy boosts and insomnia. I've recommended to every one who is in pain."

Health food store manager: " I really couldn1t stand lying on it. I had to pretend I was trying to survive torture to save my life. However, after nine days when I went to bed late I realized that I was uncomfortable until I pulled out the Panacea. It had become something that really enhanced my sleep"

Summaries of protocols of clinical trials

Effects of IPLIKATOR of Kuznetsov, courtesy of V.I. Poukov, Moscow clinic "Therapevt", personal communication, 1989

Institution: State Institute of Experimental Surgery
Location: Moscow

Date: March 9, 1982

Sample: 75 patients, 750 treatments

Results:

• Pain and stiffness were improved in all patients with osteochondrosis

• Respiratory functioning was improved in all patients with broncho-pulmonal pathology

• Normal ECG and arterial blood pressure remained unchanged during treatments

Institution: Central State Institute of Traumatology and Orthopedics

Location: Moscow

Date: March 8, 1983

Sample: 176 patients, 1500 treatments

Results:

• Highly reproducible pain relief in patients with complicated fractures of upper and lower extremities and with concussions

• Significant improvement in vision condition (decreases in double vision, decreases in blurred vision, and increases in acuity)

• No adverse side effects reported

Institution: State Institute of Physical Education

Location: Moscow

Date: December 5, 1981

Sample: patients, 2450 treatments

Results:

• Heart rate of healthy young males increased for an average of 4.5 minutes then returned to normal for the rest of the session.

• Arterial blood pressure deviated from the baseline by 10 mm Hg, increasing or decreasing, during the session.

• Oxygen consumption increased by an average 20 % during the session, gradually returning to the baseline following the session.

- CO2 expiration rate increased by an average of 17.5%, demonstrating dynamics similar to those for oxygen consumption
- Physical exercise endurance with intensity 60% of maximal oxygen consumption was increased by an average of 10% after a 30-minute duration session.
- Electrical parameters of acupoints corresponded to the endocrine glands, after the session revealed significant improvement in hormonal status of the patient.
- Skin bioluminescent radiation (effect of Kirlian) increased after the session indicating an improved functional state

Institution: State Institute of Neurosurgery

Location: Moscow

Sample: 30 patients, 1000 treatments

Results:

- A remission, observed during the acute phase of different neurological diseases, has been shown to be longer lasting, than with standard medication methods of treatment.
- The authors suggested that the mechanism of the applicator's effect is based on overall improvement in unspecific afferent flow to the pathologically changed brain.

Institution: Moscow State Medical School #2

Location: Moscow

Sample: 120 patients

Results:

- Negative phase of T-wave of ECG decreased and ST- wave flattened during the applicator session, indicating improvement in blood supply of cardiac muscle
- Tachisystolic type of cardiac arrhythmia changed to more favorable normosystolic

• Biochemical parameters of the blood demonstrated significant improvement.

Disclaimer

The information you receive from Dr. Zilberter is not a substitute for medical treatment. For any medical problem it is important that you have seen your physician and have had any medical treatment completed or underway. Dr Zilberter shall not be liable for any direct, indirect, incidental, special or consequential impairments, resulting from the inability to properly use the information or resulting from unauthorized use of the information including but not limited to, substituting the method for professional medical help.

APPENDIX

Physiological effects of endogenous endorphins

Endorphins are our candidates for explanation of the chemistry of Panacea effects. The kind of stilulation the Panacea provides and the effects that follow allowed us to put two and two together and the primal suspect was the group of brain chemicals that naturally is being produced to regulate many body functions.

Endorphins (abbreviation from endo-genous mo-rphine) were first discovered as a special class of opiate-like substances (the class includes also enkephalins and dynorphins) released in the brain and pituitary gland to take care of pain perception (and euphoria as well, but that was first considered a side effect).

The Releasers
The first discovered as well as the most classic effect of Endorphins is pain relief. Now, their functions are better known and the causes of their release are well researched.

Endorphins were first discovered in regards to their reflective pain-killing action, because they had shown to be release as an answer to pain, including chronic pain. Since than, it has been determined the most well-recognized method to release endogenous Endorphins is acupuncture including electroacupuncture as well as electrical nerve stimulation. Later researchers came to the conclusion, that any type of peripheral skin stimulation, provided with intensity sufficient to decrease pain, can be releasers.

To generalize the reasons why Endorphins are released by the body, we can say that their main purpose is to help the body withstand and overcome mild to severe exertion. Doesn't it sound suspiciously comprehensive? Just about anything can cause Endorphin release and about anything can be affected by Endorphins. However, we can be at lease sure that both Kuznetsovís and Buteyko are modalities fall within these categories: one due to intensive skin stimulation, another due to hypercapnia, they both cause release of Endorphins.

Endorphin-induced release of regulatory peptides.
In our research on endorphin releasers (130, 132, all data discussed below are referred to these books' bibliography and are available upon request), we discovered an interesting state the body falls into in many cases of disturbed physiological balance. Let just about any of numerous regulatory peptides (including endorphins) deviate from its basal concentration, and a long chain of reactions

begins. Regulatory peptides are called so because they are able to change from a few to many body's functions. For example, after endorphin concentration is increased in the body, the following peptides' concentration will change change:

The arrows designate excitatory and inhibitory indications, respectively. Somatostatin is chosen as an example of initiation of the second order cascade. Peptides in the overlapping ellipses are under influence of both inductors.

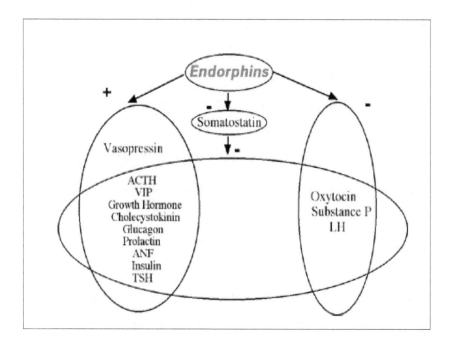

Abbreviations: ACTH adrenocorticotropic hormone; ANF atrionatriuretic factor; LH luteotropin; TSH thyreostimulating hormone; VIP vasoactive intestinal polypeptide;

Table on the next page shows the chain of events starting from stimuli releasing endorphins to the cascade of regulators to body functions under the control of these regulators.

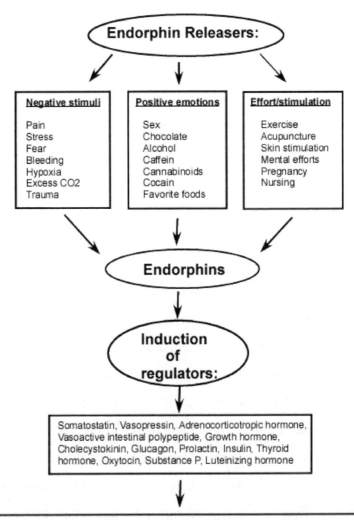

As a result, the following physiological functions will change, too: body temperature, blood pressure, skeletal muscle tone, breathing, heart rate, water-sodium balance, appetite, thirst, protein synthesis, lipolysis, hormonal balance, sexual motivation, peristalsis, immune activity, general alertness, pain perception, emotionality, and learning ability. To explain how it happens, let us take a closer look at the bodyís functions known to be under control of peptide cholecystokinin (which is just one of the many under influence of endorphins).

This peptide is rather well known due to its prominent appetite suppressive properties, but besides, it causes changes in general alertness, emotionality, motor drive, pain perception, exploratory activity, and learning ability. It also stimulates release of at least 6 other peptides: corticotrophin (ACTH), beta-endorphin, vasopressin, vasoactive intestinal polipeptide (VIP), and prolactin

Every single one of these peptides is able to induce releases of a number of other peptides, and every one of all of them is a potent physiological regulator. As a result, after just two cascade steps, the grand total of physiological changes can be quite different from those due to cholecystokinin alone. Direct research data following the time course of physiological effects regarding the peptidesí pharmacodynamics in the blood (or cerebrospinal fluid, brain, and tissues) are lacking. However, there was one very impressive result of a remote effect of cholecystokinin applied as a gum patch. It caused statistically significant improvement in the condition of drug resistant schizophrenia with the latency 3 weeks, when no traces of increased cholecystokinin in the blood could be detected.

Adaptatogenesis
Endorphins act to coordinate many of physiological functions to alter the general state of the body, for example, it can happen trough inhibiting sympathetic nervous system tone (19). As in many other cases of regulation but not direct control, endorphins tend to return a physiological function to normal: they usually increase body temperature, but decrease it during fever or during menopausal hot flashes (101, 57, 53). They decrease blood

pressure, only if it is elevated, not in normals (10, 104, 86). Endorphins have been shown to fulfill universal adaptogenous role," ...in regulation of precise conformity among homeostasis, behavior, and variable environment " (88).

How Endorphins Control Body Functions

Positive reward
Endorphins may substitute for intracranial self-stimulation reward in animals experiments, proving their important role in the brain system of pleasure (53, 59). Mood enhancement, shown to correlate with endorphin level, correlated also with analgesia and suppression of food intake (110). Normal concentration of endorphins in the blood correlated with mood enhancement and euphoria, their deficit correlated with aggressiveness. endorphins replacement treated phobias (74, 8, 110).

Placebo
Endorphins has been shown to participate in placebo effects: treating subjects with endorphins antagonist naloxon eliminated placebo response in double blind control study (37).

Stress
Endorphins participate in both local and general adaptation syndromes due to stress: they provide inhibition of plasma concentration of epinephrine and cortisol and stimulate the pineal gland, causing the release of anti stress hormone melatonin (32, 39, 66, 75, 88, 93, 12, 105).

Appetite/Hunger
Numerous experimental and clinic data show that defects in the endorphin system of the body may lead to both obesity (43, 212 77, 113, 125) and anorexia (6, 60, 81). Neurophysiological

mechanisms of food intake regulation suggest (among others) the participation of a prominent appetite suppressant Cholecystokinin (7, 57) and endorphin in activities of the specific for appetite control brain areas -- ventromedial nuclei and lateral hypothalamic areas (111, 60, 55, 77, 79). Chronic endorphins administration decreases body weight, while inhibition of endorphin activity facilitates weight gain in nervous anorexia (23, 65). However, it does not seem that endorphins control long-term processes of maintaining body weight. Rather than hunger, they modulate short-term control of appetite and mechanisms of positive reward, "pleasure" (46, 65, 79, 103, 113).

Body temperature
Endorphins lead to increased peripheral vasoconstriction, inhibition of heat loss, increase of oxidative metabolism, and modification of behavioral thermoregulation resulting in an elevation of body temperature (17, 18, 25, 44, 47, 49, 114, 117, 121).

Blood glucose
Besides their stimulatory effect on insulin secretion, endorphins provide modulation of pancreatic function to control both hyper- and hypoglycemia e.g., through stimulation of glucagon release (21, 28, 32, 43, 62, 68)

Immune system
There is a growing body of evidence about strong immunostimulatory action of endorphins
(78, 90, 51, 52, 57, 63, 72, 94, 95, 100, 101, 106, 112, 115, 116, 130).

Growth hormone
Endorphins are potent stimulators of growth hormone secretion (12, 14, 68, 88, 90, 119, 128).

Other hormones

Besides the growth hormone release, endorphins control a number of other hormonal mechanisms such as those regulating reproductive functions (parturation, lactation, etc.), stress management and environmental adaptation (96, 109, 18, 33, 37, 98, 122).

Oxygen utilization

Sufficient concentration of endogenous opiates have shown to be essential for the oxygen utilization and other defensive mechanisms during hypoxia (3, 4, 40, 69, 50).

The role endorphins in diseases
(continues next page)

Disease/condition	Ref.	Comments
Parkinson's disease	54-56	End. control Dopamine release
Hot Flashes	57, 58	Significantly lower levels of plasma beta-Endorphin were found at the onset of hot f lashes as opposed to 5 - 20 minutes before
Schizophrenia	59-61	The instability of Endorphin secretion may contribute to the pathogenesis of schizophrenia. Endorphins possess anti psychotic efficacy
Phobias	62	Endorphin replacement has have positive effect
Seizures	63-66	The lack of Endorphins correlated with a higher probability of many kinds of seizures, their injections inhibited epileptiform activities in the different regions of the brain
	67	The relaxation response of Endorphins, as well as their anti epileptic action, includes the ability to release the inhibitory neuromediator GABA
Obesity	34, 68-71	Defects in the Endorphin system may lead to obesity
Anorexia	72-74	Inhibition of Endorphin activity facilitates weight gain in nervous anorexia. Defects in the Endorphin system may lead to anorexia

PMS	75, 76	PMS had been demonstrated to correlate with endogenouse opioids system disfunction
Hypertension	12, 15, 77	End decrease elevated blood pressure, but have no effects on the normotensive rats
Alcoholism	78, 79	Social drinkers showed an increase of beta-Endorphin in plasma levels which on the contary was absent in dysphoric alcoholics.
Diabetes	33, 80, 81	End have a stimulatory effect on insulin secretion

The IPLICKATOR in the Health Journal

by AN Nekrasov, director of Central Institute of Physical Education. One of IPLICKATOR developers.

Assorted plastic circles or square modules with sharp spikes appeared for sale in the Moscow shop "Zenith" in Sokolniki Park, in pharmacies, and kiosks. This journal received tons of requests to write about "IPLICKATOR of Kuznetsov".

IPLICKATOR is successfully used to relieve pain in muscles, joints and spine, for the normalization of the cardiovascular, respiratory and nervous systems, the gastrointestinal tract, as well as to restore and improve physical efficiency. Created by a group of authors on the basis of USSR Scientific Research Institute of Medical Technology, Ministry of Health of the USSR.

The IPLICKATOR consists of multiple sharp elements (needles, thorns), which, when applied to the skin of the whole body or certain of its zone, produces its beneficial effect. In acute pain, IPLICKATOR is applied to the skin in a certain area of the body (cervical, thoracic, lumbar spine, joints) and press, within the pain tolerance for 40-60 seconds. It is necessary that people contract and relax the muscles in the area. It is repeated until the pain is reduced.

In chronic disease iplikator can be attached to the body with elastic bandage or a toweland wared for hours. Contraindications are the same as that of a therapeutic massage: tumor, pustular skin disease, thrombophlebitis.

Clinical trials were conducted in the Institute of reflexo-therapy of USSR Ministry of Health, Institute of Surgery named after AV Vishnevsky, Academy of Medical Sciences, a specialized Clinical Hospital № 8 of the RSFSR Ministry of Health, Central Research Institute of Traumatology and Orthopedics, Ministry of Health, Institute of Neurosurgery NN Burdenko, Academy of Medical Sciences.

Noting the high therapeutic efficiency the treatment, experts believe that it is not a panacea and is more effective if we combine it with medicines and physiotherapy.

Since 1979 we have worked out a way of using iplikatorov to speed recovery after heavy exercise in athletes to enhance performance in the training process and just prior to launch, as well as for the prevention and treatment of injuries. Specialists VNII Physical Culture recommended the use of "iplikatsiyu" in the preparation of the USSR team in all sports.

When cardiovascular, respiratory diseases, diseases of the stomach and intestines, spinal column and headaches, we recommend plastic IPLICKATOR with the distance between the tips of spines (step) 6-8 millimeters. For pain relief in joints and muscles - 4-b millimeters. For children, as well as people-strung, with increased pain sensitivity can use IPLICKATOR with a step 2.4 mm.

The advantages of the method should include the opportunity to use them at home.

A.N. Nekrasov, director of high sports school at the Central Institute of Physical Culture, one of the authors invention. – *The Health Journal November 1988 (translated from Russian)*

IPLICKATOR against insomnia

V.S. CHUGUNOV, MD, Clinic of neuroses RSFSR Ministry of Health

For more than 10 years, virtually all doctors in our clinic were widely used IPLICKATOR in treatment of various diseases, primarily neuroses and insomnia. A wealth of experience, revealed the details of the individual using the "hedgehogs" - so we call IPLICKATOR

I'll start with the most important thing for all those who decided to use it. First, check with your doctor, do those tests and studies that he will recommend. This is important! For example, it can stimulate the growth of various tumors, including polyps, warts.

After receiving a doctor's "OK", you can safely make the iplikator your ally. We particularly recommend it to those who are engaged in mental labor, as well as representatives of the so-called sedentary occupations. We have seen that iplikator in such cases is extremely effective. He perfectly relieves nervous tension, thereby protecting the body from the damaging effects of stress. And if you lie down for about 15-20 minutes on a flat plate modules, or sitting, leaning against the back of the chair, with IPLICKATOR wrapped over the upper back and forearm, or to put it firmly on the neck and head, you will feel relaxed, as if throwing off the burden of day concerns. This is explained by the fact that, affecting a large number of needles in the zone Zakharyan-Head, IPLICKATOR increases blood supply to the brain, internal organs, tearing pain, stimulating and exciting impulses coming from chained muscle fatigue in the central nervous system.

Sleep - a very complex process, and its quality depends on how you have lived a day. If you overeat, especially at night, sort things out on a raised voice, you can not give up late in the evening television or enticing a detective, no IPLICKATOR helps you to get rid of insomnia.

Many variety of reasons in our eventful time can result in nerve strain and concomitant insomnia. And we must, simply must help themselves, to make every effort not to bring the body to breakdown, to neurosis! I am sure everyone is able to do.

Picture yourself approaching the house. Slow down, admire the clouds in the sky, or little birds perching in the bush. Exchange a couple of words with a neighbor at the door, even sit on the bench for 3-4 minutes. And you enter the apartment more sedate. Take a shower and lay down for 15-20 minutes at IPLICKATOR. Spikes are felt only the first minute or two, then the body is overfwhelmed with heat for 5-10 minutes, you can even slumber (after all, under the influence of needle in the blood are ejected endorphins - hormones of sleep).

And when you stand up, then you will feel that you can speak with your husband and children friendly, you are willing to speak kindly, have time to cook dinner without stress, to alter a lot of household chores and watch a television program. In short, you do not fall down. The evening is calm and quiet as you get ready for bed. You can lie back on the IPLICKATOR 40-60 minutes. Read the book. Keep in mind that the area of sleep is located in the center of the back-from the middle of the blades to the waist. It is on this part of the spine you must put the iIPLICKATOR.

The Health Journal November 1988 (translated from Russian)

Iplicator of Kuznetsov helps arthrosis patients
by V.S. Chugunov, MD

Simultaneously applying IPLICKATOR of Kuznetsov to the joint and spinal region relieves pain and improves general health condition.

Neuroses are often aggravated by somatic conditions accompanied with pain. This is why we neurologists are constantly concerned with methods of decreasing the aggravating conditions. For instance, persistent pain in joints as it happens in case of deforming arthrosis. Constant joint pain growing at night and during movements, crepitation, limitation of mobility in the later stages of the disease - all of this increased irritability, reduced performance, prevents live fully.

For many years we use in our clinic the IPLICKATOR of Kuznetsov for the treatment of underlying disease and to reduce pain caused by deforming arthrosis. I want to emphasize that we cannot treat the reasons of inflammatory rheumatism and arthritis, we give recommendations specifically on deforming arthritis, especially in the initial stages.

Osteoarthritis of the first degree of gravity is characterized by moderate joint pain while walking and intermittent crunching during motion. Swelling and obliteration of the contours of the joint are possible.

With osteoarthritis of the second degree of severity pain in the joint during walking is constant, sometimes it becomes very strong. More pronounced crunch, swelling and smoothed contours of the joint are observed. Active and passive movements of joints are possible, but there may be some limitation in bending or unbending.

How to apply in these cases, IPLICKATOR of Kuznetsov? In arthrosis of the knee joint, for example, it should be applied to the front and side surfaces of the joint and be fixed with elastic bandage, than moderately pressed to the body for 30-45 minutes. Elastic bandage creates an even pressure on the entire surface of the joint, which improves blood circulation in it. WHile using the IPLICKATOR , sit and try to make some flexor and extensor movements. Rest and again move the leg. Try also slow careful walking.

IPLICKATOR can be applied on any major joint, while following the rule: the needle should not put press the neurovascular bundles, for example, in the armpit or in the groin.

We advise that you purchase an IPLICKATOR, which could cover the entire spine. Simultaneously with the application of a small piece to the joint, IPLICKATOR and lie down on it for in the supine position half an hour. Interactions of autonomic centers of the spinal cord, the application enhances metabolic processes in the body, improve the nutrition of muscles and joints, and leads to a significant improvement in general condition especially when used regularly.

We usually use applications for 10-14 days, then make a break for the same time. If you do the procedure every other day, the whole course will take a month. Then again, you'll need a break for a month.

In the case of arthrosis of the third and fourth degrees, of severity pain is very high. If in the initial stages of the disease (I and II severity) IPLICKATOR of Kuznetsov promotes a significant reduction of all major forms of osteoarthritis, in more advanced stages (III and IV degrees) it removes only the pain.

Contraindications to the use of "Iplikatora Kuznetsov: malignant and benign tumors, including papilloma, lipoma, birthmarks, thrombophlebitis.

The Health Journal, January 1989, translated from Russian

The bed of nails in Sweden

Shaktimattan is registered in the Guiness Book of Records.
In August 2009, a major manufacturer of mats with nails, Shakti, has pulled together 3 thousand people in a park in Stockholm to lie down on mats laid out to form the rays of the sun. Most of them wefre chanting mantras, generally "om Shanti Om", others went to sleep, reflecting perhaps the relaxing properties of the object. The Guiness record was registered: "Most People Chanting!"

The enormous success of the bed of nails, the Shakti Mat, or Shaktimattan in Swedish, made headlines of mass media all over the globe. The sales in 2009 reached 300 000 in Sweden alone and currently the Shakti mat is taking the world by storm. The marketing strategy is brilliant, the performance is elegant.

In March 2008, I was invited to a press conference organized by the Shakti team in Stockholm. The results were, well, interesting. Some Swedish papers wrote about the press conference, some even cited my answers to the questions fairly accuratly, for instance, Aftonbladet. But there has been also a "side effect."

On 9 July, **The organization the Consumer Representatives** (Consumer Ombudsman) sent a letter to the Team Shakti objecting the design of the company's marketing materials. *"We have adjusted our marketing based on the Consumer Ombudsman's comments and recommendations,* "says Dan Engman CEO of Team Shakti. *"The research that the team Shakti used for its current marketing materials is performed by Tanya Zilberter, who worked at the Community Holistic Health Center, Carrborough, NC. Research report can be viewed in its entirety on http://reflexo-therapy.com. This research has recently been called into question why*

Team Shakti on consumer representative's recommendation, decided to remove references to this research until new and independent research in this field that evaluates spikmattors health effects are available. Team Shakti has initiated cooperation with the new research project, where Shakti mat will be used, which will begin in the autumn. The aim is to further develop matte nail treatments and gain new clinical studies to refer to. "

In a recent commentary, the newspaper **Svenska Dagbladet** wrote that there was *"nothing that even approaches a scientific proof for the effects"* of the nail mat. I tried but failed finding on Internet or in Swedish papers any critique of the endorphin hypothesis about possible mechanisms of beneficial effects of "the bed of nails". Martin Ingvar, professor at Karolinska Institute, has criticizing those who market bed of nails for "fast cash". Dan Engman, president of the company, which launched the most popular version of the bed of nails in Sweden, argued:

"The researcher's task is to measure things and we welcome a scientific research on the effects of bed of nails." To the question: Does the bed of nails makes you healthy. Dan answered: "No, you cannot say outright, but many feel better. Something positive happens, but you do not know what. We are confident that future research may show it, "he says. (HelaHälsingland, July 09, 2009)

Opinions, especially if they don't follow the rules of scientific discussion, doesn't bother me. What does bother me is that the real inventor of the modern bed of nails, Ivan Kuznetsov is never mentioned. However, Russian origin of the Shakti mat was mentioned, though very uncertainly: *"The origins of Sweden's nail mat fad are murky, but what is clear is that it began in the yoga community and later moved into the general population. One of those credited with popularizing the mat is Susanna Lindelöw, 46, who bought a mail order nail mat made in Russia in a desperate attempt to cure severe lower back pain."* (Seeking Uncommon Sense" JANUARY 6, 2010)

Comparing the original Applicator by Kuznetsov with the current model sold as "Shakti mat" in Sweden

The Kuznetsov's unit in 1970s looked like a plastic button with several pointed needles and little holes so people could sew it on to

Kuznetsov's applicator 1970s

pieces of fabric. It looks exactly the same now in 2010. The Shakti unit, from the very beginning of its victorious march, looked completely identical to the Kuznetsov's applicator, here is how it looks in 2010:

Shakt mat, 2010

"I wandered down from the mountain and created a bed of nails"

Says Om Mokshanad:*"After a long time spent in the Himalayas practicing mysticism, yoga and meditation I found that nothing is complete until all fellow human beings can experience living in deep happiness. During the past few years I have exclusively been focusing on different yoga practices and meditations. I wandered down from the mountain and created a bed of nails with the help of the ancient Indian knowledge, Vedic science which was recorded 5-7000 years ago."* Gaia Pulse, Mar 5, 2009.

Sadly, but the actual inventor Ivan Kuznetsov, whose applicator was copied 1:1 for the Shakti mat, was mentioned by Om Mokshanad.

I would like to finish this little book on an optimistic note. The idea of Indian mystics, Ivan Kuznetsov, and the brilliant marketing solution of the Shakti team are helping people all over the world. I hope that this humble book will make its input to better understanding of the processes in our bodies. As a neuroscientist, I know very well that any type of awareness, be it the metaphors of the induistic mysticism or hypotheses based on "straight" neuroscience, contribute significantly to the body's innate ability to heal itself.

References

1. Agmo, A., Berebfeld, R. Reinforcing properties of ejaculation in the male rat : role of opioids and dopamine. Behavioral Neuroscience. 104(1) :177-82, 1990

2. Ahmed, M.S., Cemerikis, B., Agbas, A. Properties and functions of human placental opioid system. Life Sciences. 50 (2) : 83-97, 1992.

3. Akiyama, Y. Role of endogenous opioids in respiratory control system and dyspnea sensation in healthy adult humans. Hokkaido Igaku Zasshi-Hokkaido Journal of Medical Science. 67(1); 131-40, 1992

4. Akiyama, Y., Nishimura, M., Suzuki, A., Yamamoto, M., Kishi, F., Kawakami, Y. Naloxone increases ventilatory

response to hypercapnic hypoxia in healthy adult humans. American Review of Respiratory Disease. 142(2) :301-5, 1990

5. Bajorek, J.G., Lee, R.J., Lomax, P. Neuropeptides: anticonvulsant and convulsant mechanisms in epileptic model system and in humans. Advances in Neurology. 44 :489-500, 1986.

6. Baranowska, B. Are disturbances in opioid and adrenergic system involved in the hormonal dysfunction of anorexia nervosa? Psychoneuroendocrinology. 15(5-6) :371-9, 1990.

7. Benoliel, J.J., Mauborgne, A., Ledrang, Jr., Hamon, M., Cesselin, F. Opioid control of the in vitro release of cholecystokinin-like material from the rat substantia nigra. Journal of Neurochemistry. 58(3) : 916-22, 1992

8. Bergmann, F. The role of endogenous opioid peptides in physiological and pharmacological reward responses--a survey of present-day knowledge. Israel Journal of Medical Sciences. 23(I-2) : 8-11, 1987

9. Berlowitz, D., Denehy, L., Johns, D.P., Bish, R.M., Walters, E/H.
The Buteyko asthma breathing technique. Medical Journal of Australia. 162(1):53, 1995

10. Bjorntorp, P. Effects of physical training on blood pressure in hypertension. European Heart Journal. 8 Suppl B :71-6, 1987

11. Borer, K.T., Nicoski, D.R., Owens, V. Alteration of pulsatile growth hormone secretion by growth-inducing exercise: involvement of endogenous opiates and somatostatin. Endocrinology. 118(2) :844-50, 1986

12. Brown, R., King, Mg., Husband, A.J. Sleep deprivation-induced hyperthermia following antigen challenge due to opioid but not interleukin-1 involvement. Physiology & Behavior. 51(4) :767-70, 1992

13. Bruhn, .TO., Tresco, P.A., Mueller, G.P., Jackson, I.M. Beta-endorphin mediates clonidine stimulated growth hormone release. Neuroendocrinology. 50(4) :460-3, 1989

14. Brunetti, L. Molecular basis of the communication of the nervous and immune system. Clinica Terapeutica. 136(5) :301-20, 1991

15. Bullard, D.E. Diencephalic seizures: responsiveness to bromocriptine and morphine. Annals of Neurology. 21(6) : 609-11, 1987

16. Cagampang, F.R., Maeda, K.I., Tsukamura, H., Ohkura, S., Ota, K. Involvement of ovarian steroids and endogenous opioids in the fasting-induced suppression of pulsatile LH release in ovariectomized rats. Journal of Ehdocrinology. 129(3) : 321-8, 1991

17. Cagnacci, A., Bonuccelli, U., Mels, G.B., Soldani, R., Piccini, P., Napolitano, A., Muratorio, A., Fioretti, P. Effect of

naloxone on body temperature in postmenopausal women with Parkinson's disease. Life Sciences. 46(17):1241-7, 1990.

18. Casper, R.F. Disorders of the hypothalamic pulse generator : insufficient or inappropriate gonadotropin-releasing hormone release. Clinical Obstetrics & Gynecology. 33(3) : 611-21, 1990

19. Casper, R.F., Yen, S.S. Neuroendocrinology of menopausal flushes: an hypotesis of flush mechnism. Apr 12 18:28 1995

20. Champeroux, P., Brisac, A.M., Laurent, S., Schmitt, H. Endogenous opiate system and dihydropyridine-induced central regulation of sympathetic tone in rats. European Journal of Pharmacology. 158(1-2) : 157-60, 1988

21. Chiodera, P., Coiro, V. Endogenous opioid mediation of somatostatin inhibition of arginine vasopressin release evoked by insulin-induced hypoglykemia in man. Journal of Neural Transmission-General Section. 83(1-2) : 121-6, 1991.

22. Cochen, M.R., Pickar, D., Cochen, R.M., Wise, T.N., Cooper, J.N. Plasma cortisol and beta-endorphin immunoreactivity in human obesity. Psychosomatic Medicine, 46(5):454-62, 1984

23. Davis, J.M., Lamb, D.R., Vim, G.K., Malven, P.V. Opioid modulation of feeding behavior following repeated exposure to forced swimming exercise in male rats. Pharmacology, Biochemistry & Bechavior. 23(5) : 709-14, 1985

24. De Marinis L. Mancini A. D,Amico C. Sambo P. Tofani A. La Brocca A. Barbarino A. (Role of endogenous opioids in the modulation of hypophyseal hormone secretion.) (Review) (Italian) Minerva Medica. 81(1-2) :5-14, 1990 Jan- Feb.

25. De Meirleir, K., Arentz, T., Hollmann, W., Vanhaelst, L. The role of endogenous opiates in thermal regulation of the body during exercise. British Medical Journal Clinical Research Ed. 290(6470) :739-40, 1985

26. Della Bella, D., Carenzi, A., Frigeni, V., Reggiani, A., Zambon, A. Involvement of monoaminergic and peptidergic components in cathinone-induced analgesia. European Journal of Pharmacology. 114(2) : 231-4, 1985

27. Dmitriev, A.D, Kizim, E.A., Smirnova, M.B., Shcheglova, I.D., Ugolev, A.M. The endorphins of the epithelial and subepithelial structures of the rat small intestine and the evolutionary hypotheses of the formation of the mechanisms of the negative regulation of their synthesis. Zhurnal Evoliutsionnoi Biokhimii i Fiziologii. 27(6) :701-11, 1991

28. Dooley, C.P., Saad, C., Valenzuel,a J.E. Studies of the role of opioids in control of human pancreatic secretion. Digestive Diseases & Sciences. 33(5) :598-604, 1988

29. Drake, C.T., Terman, G.W., Simmons, M.L., Milner, T.A., Kunkel, D.D., Schwartzkroin, P.A., Chavkin, C. Dynorphin opioids present in dentate granule cells may function as retrograde inhibitory neurotransmitters. Journal of Neuroscience. 14(6) : 3736-50, 1994

30. Duarte, I.D., Ferreira-Alves, D.L., Nakamura-Craig, M. Possible participation of endogenous opioid peptides on the mechanism involved in analgesia induced by vouacapan. Life Sciences. 50(12) : 891-7, 1992.

31. Duggan, A.W. Pharmacology of descending control systems. Philosophical Transactions of the Royal Sotiety of London- Series B: Biologycal Sciences. 308(1136) :375-91, 1985

32. el-Tayeb, K.M., Brubaker, P.L., Cook, E., Vranic, M. Effect of opiate-receptor blockade on normoglycemic and hypoglycemic glucoregulation. American Journal of Physiology. 250 (3 Pt 1): E236-42, 1986

33. Endo, Y., Jinnai, K., Kimura, F. The roles of endogenous opioids in the inhibitory action of the hippocampus on preovulatory luteinizing hormone in rats. Endocrinologia Japonica. 37(4) : 535-43, 1990

34. Eriksson, S.V., Lundeberg, T., Lundeberg, S. Interaction of diazepam and naloxone on acupuncture induced pain relief. American Journal of Chinese Medicine. 19(1) :1-7, 1991.

35. Facchinetti ,F., Martignoni, E., Sola, D., Petraglia, F., Nappi, G., Genazzani, A.R. Transient failure of central opioid tonus and premenstrual symptoms. Journal of Reproductive Medicine. 33(7) :633-8, 1988

36. Facchinetti, F., Genazzani, A.D., Martignoni, E., Fiorini, L., Sances, G., Genazzani, A.R. Neuroendocrine correlates of premenstrual syndrom: changes in the pulsatile pattern of plasma LH. Psychoneuroendocrinology. 15(4):269-77, 1990.

37. Farrell, P.A., Gustafson, A.B., Garthwaite, T.L., Kalkhoff, R.K., Cowley, A.W. Jr., Morgan, W.P. Influence of endogenous opioids on the response of selected hormones to exercise in humans. Jourmal of Applied Physiology. 61(3): 1051-7, 1986

38. Fiatarone, M.A., Morley, J.E., Bloom, E.T., Benton, D., Makinodan, T., Solomon, G.F. Endogenous opioids and the exercise-induced augmentation of natural killer cell activity Journal of Laboratory & Clinical Medicine. 112(5):544-52, 1988

39. Fuenmayor, N., Cubeddu, L. Cardiovascular and endocrine effects of naloxone compared in normotensive and hypertensive patients. European Journal of Pharmacology. 126(3):189-97, 1986

40. Gamble, G.D., Milne, R.J. Hypercapnia depresses nociception: endogenous opioids implicated. Brain Research. 514(2);198-205, 1990

41. Garris, P.A., Ben-Jonathan, N. Regulation of dopamine in vitro from the posterior pituitary by opioid peptides. Neuroendocrinology. 52(4): 399-404, 1990 Oct.

42. Gil-Ad, I., Dickerman, Z., Amdursky, S., Laron, Z. Diurnal rhythm of plasma beta endorphin, cortisol and growth hormone in

schizophrenics as compared to control subjects. Psychopharmacology. 88(4) :496-9, 1986.

43. Giugliano, D., Salvatore, T., Cozzolino, D., Ceriello, A,. Torella, R., D'Onofrio, F. Hyperglycemia and obesity as determinants of glucose, insulin, and glucagon responses to beta-endorphin in human diabetes mellitus. Journal of Clinical Endocrinology & Metabolism. 64(6) : 1122-8,1987

44. Gordon, C.J., Rezvani, A.H., Heath, J.E. Role of beta-endorphin in the control of body temperature in tha rabbit. Neuroscience & Biobehavioral Reviews. 8(1) :73-82, 1984 Spring.

45. Grossman, A., Sutton, J.R. Endorphins: what are they? How are they measured? What is their role in exercise?. Medicine & Science in Sports & Exercise. 17(1) :74-81, 1985

46. Gunion, M.W., Peters, R.H. Pituitary beta-endorphin, naloxone, and feeding in several experimental obesities. American Journal of Phsiology. 241(3) :R173-84, 1981

47. Gwosdow, A.R., Besch, E.L. Adrenal and thyroid interactions of beta-endorphin-induced body temperature responses of rats at 24,5 degfrees C. Proceedings of the Sotiety for Experimental Biology & Medicine. i78(3) :412-8, 1985

48. Han, J.S., Zhang, R.L. Suppression of morphine abstinence syndrome by body electroacupuncture of different frequencies in rats. Drug & Alcohol Dependence. 31 (2) : 169-75, 1993

49. Handler, C.M., Geller, E.B., Adler, M.W. Effect of mu-, kappa-, and delta-selective opioid agonists on thermoregulation in the rat. Pharmacology, Biochemistry & Behavior. 43(4):1209-16, 1992

50. Harber, V.J., Sutton, J.R., Endorphins and exercise. Sports Medicine. 1(2): 154-71, 1984

51. Harbour, D.V., Galin, F.S., Hughes, T.K., Smith, E.M., Blalock, J.E. Role of leukocyte-derived pro-opiomelanocortin peptides in endotoxic shock. Circulatori Shock. 35(3):181-91. 1991

52. Heijnen, C.J., Kavelaars, A. The contribution of neuroendocrine sudstances to the immune response. Netherlands Journal of Medicine. 39(3-4):281-94, 1991

53. Hinton, E.R., Taylor, S. Does placebo response mediate runner's high?. Perceptual & Motor Skills. 62(3):789-90, 1986

54. Ho, S.B., DeMaster, E.G., Shafer, R.B., Levine, A.S., Morley, J.E., Go, V.L., Allen, J.I. Opiate antagonist nalmefene inhibits ethanol-induced flushing in Asians: a preliminary study. Alcoholism, Clinical & Experimental Research. 12(5):705-12, 1988 Oct.

55. Hoebel, B.G. Brain neurotransmitters in food and drug reward. [Review]. American Journal of Clinical Nutrition. 42(5 Suppl): 1133-50, 1985

56. Introini-Collison, I.B., Baratti, C.M. Opioid peptidergic system modulate the activity of beta-adrenergic mechanisms during memory consolidation processes. Behavioral & Neural Biology. 46(2) :227-41, 1986

57. Jankovic, B.D., Radulovic, J. Enkephalins, brain and immunity: modulation of immune responses by methionine-enkephalin injected into the cerebral cavity. International Journal of Neuroscience. 67(1-4) :241-70, 1992

58. Kapas, L., Benedek, G., Penke, B. Cholecystokinin interferes with the thermoregulatory effect of exogenous and endogenous opioids. Neuropeptides. 14(2) :85-92, 1989

59. Katama, K., Yoshida, S., Kameyama, T. Antagonism of footshock stress-induced inhibition of intracranial self-stimulation by naloxone or methamphetamine. Brain Research. 317(1) : 197-200, 1986

60. Kaye, W,H., Pickar, D., Naber, D., Ebert, M.H. Cerebrospinal fluid opioid activity in anorexia nervosa. American Journal of Psychiatry. 139(5) :643-5, 1982

61. Kehoe, P., Blass, E.M. Behaviorally functional opioid system in infant rats: 11. Evidence for pharmacological, physiological, and psychological mediation of pain and stress. Behavioral Neuroscience. 100 (5) : 624-30, 1986

62. Khawaja, X.Z., Bailey, C.J., Green, I.C. Central mu, delta, and kappa opioid binding sites, and brain and pituitary beta-

endorphin and met-enkephalinin genetically obese (ob/ob) and lean mice. Life Sciences. 44(16) : 1097-105, 1989.

63. Kita, T., Kikuchi, Y., Oomori, K., Nagata, I. Effects of opioid peptides on the tumoricidal activity of spleen cells from nude mice with or without tumors. Cancer Detection & Prevention. 16(4) :211-4, 1992.

64. Kiyatkin, E.A. Nociceptive sensitivity/behavioral reactivity regulation in rats during aversive states of different nature: its mediation by opioid peptides. International Journal of Neuroscience. 44(1-2) :91-110, 1989

65. Levine, A.S., Morley, J.E., Gosnell, B,A., Billington, C.J., Bartness, T.J. Opioids and consummatory behavior. Brain Research Bulltin. 14(6) :663-72, 1985

66. Lissoni, P., Esposti, D., Esposti, G., Mauri, R., Resentini, M., Morabito, F., Fumagalli, P., Santogostino, A., Delitala, G., Frashini, F. A clinical study on the relationship between the pineal gland and the opioid system. Journal of Neural Transmission. 65(2) : 63-73, 1986.

67. Llewe, G., Schneider, U., Krause, U., Beyer, J. Naloxone increases the response of growth hormone and prolactin to stimuli in obese humans. Juornal of Endocrinological Investigation. 10(2) :137-41, 1987

68. Locatelli, A., Spotti, D., Caviezel, F. The regulation of insulin and glucogon secretion by opiates: a study with naloxone in healthy humans. Acta Diabetologica Latina. 22(1): 25-31, 1985

69. Lou, H.C., Tweed, W.A., Davis, J.M. Endogenous opioids may protect the perinatal brain in hypoxia. Developmental Pharmacology & Therapeutics. 13(2-4):129-33, 1989.

70. Louisy, F., Guezennec, C.Y., Lartigue, M., Aldigier, J.C., Galen, F.X. Influence of endogenous opioids on atrial natriuretic factor release during exercise in man.

71. Luu, M., Boureau, F. Acupuncture in pain therapy : current concepts. Therapeutishe Umschau. 46(8) : 518-25, 1989

72. Lysle, D.T., Luecken, L.J., Maslonek, K.A. Modulation of immune status by a conditioned aversive stimulus: evidence for the involvement of endogenous opoids. Brain, behavior, & Immunity. 6(2) :179-88, 1992

73. Maestroni, G.J., Conti ,A. The pineal neurohormone melatonin stimulates activated CD4+, Thy-1+cells to release opioid agonist (s) with immunoenhancing and anti-stress properties. Journal of Neuroimmunology. 28 (2) : 167-76, 1990

74. Martensz, N.D., Vellucci, S.V., Kenerve, E.B., Herbert, J. Beta-Endorphin levels in the cerebrospinal fluid of male talapoin monkeys in social groups related to dominance status and the luteinizing hormone response to naxone. Neuroscience. 18(3) ; 651-8,1986

75. McCubbin, J.A., Kaplan, J.R., Manuck, S.B., Adams, M.R. Opioidergic inhibition of circulatory and endocrine stress responses in cynomolgus monkeys: a preliminary study. Psychosomatic Medicine. 55(1) :23-8 1993

76. McGinty, J.F., Kanamatsu, T., Obie, J., Hong, J.S. Modulation of opioid peptide metabolism by seizures: differentiation of opioid subclasses. NIDA Research Monograph Series. 71 :89-101, 1986.

77. McLaughlin, C.L., Baile, C.A., Della-Fera, M.A. Meal-stimulated increased concentrations of beta-endorphin in the hypothalamus of Zuker obese and lean rats. Physiology & Bechavior. 35(6) : 891-6, 1985

78. Mediratta, P.K., Das, N., Gupta, V.S., Sen, P. Modulation of humoral immune responses by endogenous opioids Journal of Allergy & Clinical Immunology. 81(1) :27-32, 1988

79. Millan, M.J., Czlonkowski, A., Pilcher, C.W., Almeida, O.F., Millan, M.N., Colpaert, F.C., Herz, A. A model of chronic pain in the rat: functional correlates of alterationsin the activity of opioid systems. Journal of Neuroscience. 7(1) : 77-87, 1987

80. Millan, M.J., Millan, M.N., Reid, L.D., Herz, A. The role of the mediobasal arcuate hypothalamus in relation to opioid systems in the control of ingestive behavior in the rat. Brain Reserch. 381(1) :29-42, 1986

81. Mills, I.H. The neuronal basis of compulsive behavior in anorexia nervosa. Journal of Psychiatric Research. 19(2-3):231-5, 1985

82. Mitchell, J.B., Gratton, A. Opioid modulation and sensitization of dopamine release elicited by sexually relevant stimuli : a high speed in freely behaving rats. Brain Research. 551 (1-2), 1991

83. Mooring, F.J., Hughes, G.S. Jr., Johnsrude, I.S. Role of beta-endorphins in analgesia associated with reactrive hyperemia. Investigative Radiology. 20(3) : 293-6, 1985

84. Morel, G., Pelletier, G. Endorphinic neurons are contacting the tuberoinfundibular dopaminergic neurons in the rat brain. Peptides. 7(6) : 1197-9, 1986

85. Morgan, W.P. Affective beneficence of vigorous physical activiti. Medicine & Science in Sport & Exercise. 17(1) :94-100. 1985

86. Mosqueda-Garcia, R., Eskay, R., Zamir, N., Palkovits, M., Kunos G. Opioid-mediated cardiovascular effects of clonidine in spontaneously hypertensive rats: elimination by neonatal treatment with monosodium glutamate. Endocrinology. 118(5) :1814-22, 1986

87. Murakami, Y., Kato, Y., Koshiyama, H., Inoue, T., Ishikawa, Y., Imura ,H. Involvement of alpha-adrenergic and GABAergic

mechanisms in growth hormone secretion induced by central somatostatin in rats. Brain Research. 407(2) :405-8,

88. Nieber, K., Oehme, P. Stress and the endogenous opioid system. III. Classification of the opioid system in the process of adaptation. Zeitschrift fur die Gesamte Innere Madizin und Ihre Grenzgebiete.40(5) :133-6, 1985

89. Okajiama, S., Okajima, T., Kato, K., Ibayashi, H. Effect of taurine on growth hormone and prolactin secretion in rats: possible interaction with opioid peptidergic system. Life Sciences. 43(10) :807-12, 1988.

90. Oleson, D.R., Johnson, D.R. Regulation of human natural cytotoxicity by enkephalins and selective opiate agonists. Brain, Behavior, & Immunity. 2(3) :171-86, 1988

91. Otake, K., Kondo, K., Oiso, Y. Possible involvement of endogenous opioid peptides in the arginine vasopressin release by gamma-aminobutyric acid in conscious rats. Neuroendocrinology. 54(2) :170-4, 1991

92. Paredes, R.G., Manero, M.C., Haller, A.E., Alvarado, R., Agmo, A. Sexual behavior enhances postictal depression in kindled rats: opioid involvement. Behavioral Brain Reserch. 52(2) :175-82, 1992

93. Parrot, R.F., Thornton, S.N. Opioid influences on pituitary function in sheep under basal conditions and during psychological stress. Psychoneuroendocrinology. 14(6) ;451-9, 1989.

94. Pasotti, D., Mazzone, A., Lecchini, S., Frigo, G.M., Ricevuti, G. The effect of opioid peptides on periphenal blood granulocytes. Rivista Europea Per Le Science Medicine e Farmacologiche. 15(2) :71-81, 1993

95. Pedersen, B.K., Tvede, N. The immune system and physical training. Ugeskrift for Laeger. 155(12) :856-62, 1993

96. Petraglia, F., Vale, W., Rivier, C. Beta- endorphin and dynorphin participate in the stress-induced release of prolactin in the rat. Neuroendocrinology. 45(5) ; 338-42, 1987

97. Petrozzino, J.J., Scardella, A.T, Santiago, T.V., Edelman, N.H. Dichloroacetate blocks endogenous opioid effects during inspiratory flow-resistive loading. Journal of Applied Phisiology. 72(2) : 590-6, 1992

98. Preiffer, A., Herz, A. Endocrine actions of opioids. Hormone & Metabolic research. 16(8) :386-97, 1984

99. Przewlocka, B., Sumova, A., Lason, W. The influence of conditioned fear-induced stress on the opioid system in the rat. Pharmacology, Biochemistry & Behavior. 37(4) : 661-6, 1990

100. Przewlocki, R., Hassan, A.N., Lason, W., Epplen, C., Herz, A., Stein, C. Gene expression and localization of opioid peptides in immune cells of inflamed tissue : functional role in antinociception. Neuroscience. 48(2) :491-500, 1992.

101. Ray, A., Mediratta, P.K., Sen, P. Modulation by naltrexone of stress-induced changes in humoral immune responsiveness and gastric mucosal integrity in rats. Physiology & Behavior. 51(2) :293-6, 1992

102. Rebar, R.W., Spitzer, I.B. The physiology and measurement of hot flushes. American Journal of Obstetrics & Gynecology. 156(5) : 1284-8, 1987

103. Robert, J.J., Orosco, M., Rouch, C., Jacquot, C., Cohen, Y. Effects of opiate agonists and an antagonist on food intake and brain neurotransmitters in normophagic and obese "cafeteria" rats. Pharmacology, Biochemistry & Bechavior. 34(3) :577-83, 1989

104. Rosella-Dampman, L.M., Emmert, S.E., Keil, L.C., Summy-Long, J.Y. Differential effects of naloxone on the release of neurohypophysial hormones in normotensive and spontaneously hypertensive rats. Brain Research. 325(1-2) :205-14, 1985

105. Sagrillo, C.A., Voogt ,J.L. Endogenous opioids mediate the nocturnal prolactin surge in the pregnant rat. Endocrinology. 129 (2) : 925-30, 1991

106. Schafer, M., Carter, L., Stein, C. Interleukin 1 beta and corticotropin-releasing factor inhibit pain by releasing opioids from immune cells in inflamed tissue. Proceedings of the National Academy of Science of the United States of America. 91(10) : 4219-23, 1994

107. Schmauss, C., Emrich, H.M. Dopamine and the action of opiates: a reevaluation of the dopamine hypothesis of schizophrenia. With special consideration of the role of endogenous opioids in the pathogenesis of schizophrenia. Biological Psychiatry. 20 (11) ;1211-31, 1985

108. Segal, B.S., Inman, J.D., Moss, I.R. Respiratory responses of piglets to hypercapnia during postnatal development: effects of opioids. Pediatric Pulmonology. 11(2) ;113-9, 1991.

109. Selmanoff, M., Gregerson, K.A. Suckling-induced prolactin release is suppressed by naxolone and simulated by beta-endorphin. Neuroendocrinology. 42(3) ;255-9, 1986.

110. Sforzo, G.A., Opioids and exercise. An update. Sports Medicine. 7(2) :109-24, 1989

111. Sikdar, S.K., Oomura, Y. Selective inhibition of glucose-sensitive neurons in rat lateral hypothalamus by noxious stimuli and moephine. Journal of Neurophysiology. 53(1) : 17-31, 1985

112. Sirinek, LP., O,Dorisio, M.S. Modulation of immune function by intestinal neuropeptides. Acta Oncologica. 30(4) : 509-17, 1991.

113. Slavo, D., Facchinetti, F., Barletta, C., Petraglia, F., Buzzetti, R., Monaco, M., Giovannini, C., Genazzani, R. Plasma beta- endorphin in response to oral glucose tolerants test in obese patients. Hormone & Metabolic Research. 19(5) :204-7, 1987

114. Smith, F.L., Welch, S.P., Dombrowski, D.S., Dewey, W.L. The role of endogenous opioids as mediators of the hypothermic effects of intrathecally administered calcium and calcitonin gene-related peptide in mice Journal of Pharmacology & Experimental Therapeutics. 266(3) : 1407-15, 1993

115. Stefano, G.B. Invertebrate and vertebrate neuroimmune and autoimmunoregulatory commonalties involving opioid peptides. Cellular & Molecular Neurobiology. 12(5) :357-66, 1992

116. Stein, C. Neuro-immune interactions in pain. Critical Care Medicine. 21(9 suppl) :357-8, 1993

117. Tepper, R., Neri, A., Kaufman, H., Schoenfeld, A., Ovadia, J. Menopausal hot flushes and plasma beta-endorphins. Obstetrics & Gynecology. 70(2) :150-2, 1987

118. Thompson, D.L., Weltman, J.Y., Rogol, A.D., Metzger, D.L., Veldhuis, J.D., Weltman, A. Cholinergic and opioid involvement in release of growth hormone during exercise and recovery. Journal of Applied Physiology. 75(2) : 870-8, 1993

119. Thompson, M.L., Miczek, K.A., Noda, K., Shuster, L., Kumar, N.S. Analgesia in defeated mice: evidence for mediation via central rather than pituitary or adrenal endogenous opioid peptides. Pharmacology, Biochemistry & Behavior. 29(3) ;451-6, 1988

120. Thornhill, J.A., Saunders, W.S. Thermoregulatory (core, surface and metabolic) responses of unrestrained rats to repeated POAH injections of beta-endorphin or adrenocorticotropin. Peptides. 5(4) :713-9, 1984

121. Tseng, L.F., Li, C.H. Beta- endorphin : hyperthrmia in mice by intravenous injection. International Lournal of Peptide & Protein Research. 15(5) :471-4, 1980

122. Van de Heijning, B.J., Koekkoek-Van den Herik, I., Maigret, C., Van Wimersma Greidanus, T.B. Pharmacological assessment of the site of action of opioids on the release of vasopressin and oxytonic in the rat. European Jurnal of Pharmacology. 197(2) : 175-80

123. Van Leeuwen, A.F., Evans, R.G., Ludbrook, J. Haemodynamic responses to acute blood loss : new roles for the heart, brain and endogenouse opioids. Anaesthesia & Intensve Care. 17(3) :312-9, 1989

124. Verhoeven, W.M., van Ree, J.M., Westenberg, H.G., Krul, J.M., Brouwer, G.J., Thijssen, J.H., de Praag, H.M., Ceulemans, D.L., Kahn, R.S. Clinical, biochemical, and hormonal aspects of treatment with Des-tyrl-gamma-endorphin in schizophrenia. Psychiatry Research. 11(4) :329-46, 1984

125. Vogt, T., Belluscio, D. Controversies in plastic surgery: Suction-assisted lipectomy (SAL) and the hCG (human chorionic gonadotropin) protocol for obesity treatment. Aesthetic Plastic Surgery. (3) :131-56, 1987.

126. Wang, H., Pappas, G.D. Adrenal medullary in the rat spinal cord reduce nociception in a chronic pain model. Pain. 42(1) : 69-79, 1990

127. Yashpal, K., Henry, J.L. Neural mediation of the cardiovascular responses to intrathecal administration of substance P in the rat: Slowing of the cardioacceleration by an adrenal opioid factor. Neuropeptides. 25(6) : 331-42, 1993

128. Yenehara, N., Imai, Y., Chen, J.Q., Takiuchi, S., Inoki, R. Influence of opioids on substance P release evoked by antidromic stimulation of primary afferent fibers in the hind instep of rats. Regulatory Peptides. 38(1) ; 13-22, 1992

129. Zhao, J.C., Liu, W.Q. Relationship between acupuncture-induced immunity and the regulation of central neurotransmitter system in rabbits--11. Effect of the endogenous opioid peptides on the regulation of acupuncture-induced immune reaction. Acupuncture & Electro- Therapeutics Research. 14(1) :1-7, 1989.

130. Zilberter, T.M., Loukianova, L.L. Physiological properties of regulatory peptides, Moscow, VINITI, 1991

131. Ziberter, T.M., Roman J. Reflexo-therapy with mechanical skin stimulation: pilot study. Alternative and complementary Therapies, in press (preprint is available upon request)

132. Zilberter, T.M., Titov, S.A., Loukianova, L.L. Cascade effects of regulatory peptides, Moscow, VINITI, 1989.

133. Zilberter, T.M. To breath or not to breath? Healself Network February 1997

134. Zilberter, T.M. Breathing and pH. Healself Network July 1997

Additional references including articles published in 2000s

135. Lam, H Nurmi, N Rouvinen, K Kiianmaa, C. Effects of acute ethanol on β-endorphin release in the nucleus accumbens of selectively bred lines of alcohol-preferring AA and alcohol-avoiding ANA rats. MP . Psychopharmacology, 208, 1, 121-130, 2010

136. Bruehl S, Burns JW, Chung OY, Ward P, Johnson B. Anger and pain sensitivity in chronic low back pain patients and pain-free controls: the role of endogenous opioids. Pain 2002;99:223–33.

137. Bruehl S, Chung OY, Burns JW, Biridepalli S. The association between anger expression and chronic pain intensity: evidence for partial mediation by endogenous opioid dysfunction. Pain 2003;106:317–24.

138. Cabıoglu M.. Changes in Serum Leptin and Beta Endorphin Levels with Weight Loss by Electroacupuncture and Diet Restriction in Obesity Treatment. The American Journal of Chinese Medicine, 2006, Vol: 34 Issue: 1: 1 - 11

139. Acute Pressor and Hormonal Effects of -Endorphin at High Doses in Healthy and Hypertensive Subjects: Role of Opioid Receptor Agonism. Domenico Cozzolino, Ferdinando C. Sasso, Donato Cataldo, Domenico Gruosso, Armando Giammarco, Antonella Cavalli, Cristiana Di Maggio, Giuseppe Renzo, Teresa Salvatore, Dario Giugliano, and Roberto Torella. Journal of Clinical Endocrinology & Metabolism 90(9):5167–5174

140. V.O. Fuentes, C. Villagran, J. Navarro P.I. Fuentes. Effect of small doses of naloxone on sexual exhaustion in White New Zealand male rabbits. Animal Reproduction Science 90 (2005) 341–346

141. Jorge Juárez a, *, Gabriela Camargo a, Ulises Gómez-Pinedo. Effects of estradiol valerate on voluntary alcohol consumption, β-endorphin content and neuronal population in hypothalamic arcuate nucleus. Pharmacology, Biochemistry and Behavior 85 (2006) 132–139

142. M. Jutkiewicz. Mood-Elevating Properties of DOR Agonists. Molecular Interventions, 2006, 6, 3:162-169

143. P. E. Molina. Opioids and opiates: analgesia with cardiovascular, haemodynamic and immune implications in critical illness. J Intern Med 2006; 259: 138–154

144. Krisztina Monory & Beat Lutz. Pain killer without a high. Nature Medicine, 11, 4:378-379, 2005

145. Keiichi Niikura, Minoru Narita ⋆, Michiko Narita, Atsushi Nakamura, Daiki Okutsu,Ayumi Ozeki, Kana Kurahashi, Yasuhisa Kobayashi, Masami Suzuki, Tsutomu Suzuki. Direct evidence for the involvement of endogenous -endorphin in the suppression of the morphine-induced rewarding effect under a neuropathic pain-like state. Neuroscience Letters 435 (2008) 257–262

146. Ilana Roth-Deri, Tamar Green-Sadan, Gal Yadid. b-Endorphin and drug-induced reward and reinforcement. Progress in Neurobiology 86 (2008) 1–21

147. Sadigh B, Berglund M, Fillingim RB, Sheps D, Sylvén C.. beta-Endorphin modulates adenosine provoked chest pain in men, but not in women-a comparison between patients with ischemic heart disease and healthy volunteers. Clin J Pain. 2007 Nov-Dec;23(9):750-5

148. Kulkarni, Milind Manohar, Nadaf Ramzan, Lakshmanan,Sala Bose, Rantham, Prabhakara Jagadeesh Chandra. Skin Stimulation Device and a Method and Computer Program Product for Detecting a Skin Stimulation Location. United States Patent Application 20080312718, 2008

149. Mohab M. Ibrahim, Frank Porreca,, Josephine Lai, Phillip J. Albrecht, Frank L. Rice, Alla Khodorova§, Gudarz Davar, Alexandros Makriyannis ∥, Todd W. Vanderah, Heriberto P. Mata, and T. Philip Malan, Jr. CB2 cannabinoid receptor activation produces antinociception by stimulating peripheral release of

endogenous opioids Proceedings of the National Acad Sciences, vol. 102 no. 8, 3093–3098

150. Samuel Jarjour, Li Bai, and Christina Gianoulakis. Effect of Acute Ethanol Administration on the Release of Opioid Peptides From the Midbrain Including the Ventral Tegmental AreaAlcoholism: Clinical and Experimental Research Volume 33 Issue 6, Pages 1033 - 1043, 2009

151. Michael M. Poplawski, Nadka Boyadjieva, Dipak K. Sarkar. Vasoactive Intestinal Peptide and Corticotropin-Releasing Hormone Increase β-Endorphin Release and Proopiomelanocortin Messenger RNA Levels in Primary Cultures of Hypothalamic Cells. Alcoholism: Clinical and Experimental Research Volume 29 Issue 4, Pages 648 - 655, 2006

152. J C Kaski. Cardiac syndrome X in women: the role of oestrogen deficiency. Heart 2006;92:iii5-iii9

153. Tzong-Cherng Chia, Yi-Jin Hoa, Win-Pin Chena, Tsung-Li Chic, Shoei-Sheng Leeb, Juei-Tang Chengd and Ming-Jai Sua. Serotonin enhances β-endorphin secretion to lower plasma glucose in streptozotocin-induced diabetic rats. Life Sciences Volume 80, Issue 20, 24 April 2007, Pages 1832-1838

154. M. Cavallini, A. Casat. A prospective, randomized, blind comparison between saline, calcium gluconate and diphoterine for washing skin acid injuries in rats: effects on substance P and β-endorphin release. European Journal of Anaesthesiology (2004), 21:5:389-392

155. Bacharach, Samuel B.; Bamberger, Peter A.; Doveh, Etti. Firefighters, critical incidents, and drinking to cope: The adequacy of unit-level performance resources as a source of vulnerability and protection. Journal of Applied Psychology. Vol 93(1), Jan 2008, 155-169

156. D. Tomaszewska-Zaremba, F. Przekop, K. Mateusiak, The involvement of GABA (A) receptors in the control of GnRH and betaendorphin release, and catecholaminergic activity in ventromedialinfundibular region of hypothalamus in anestrous ewes, J. Physiol. Pharmacol. 52 (2001) 489–500

157. Shaik Shavalia, Begonia Hob, Piyarat Govitrapongb, Saiphon Sawlomb, Amornpan Ajjimapornb, cSirirat Klongpanichapakb, and Manuchair Ebadi. Melatonin exerts its analgesic actions not by binding to opioid receptor subtypes but by increasing the release of β-endorphin an endogenous opioid. Brain Research Bulletin, Volume 64, Issue 6, 30, 1572005, Pages 471-479

158. A Shaheda Daniel Shoskes. Correlation of β-endorphin and prostaglandin e2 levels in prostatic fluid of patients with chronic prostatitis with diagnosis and treatment response. The Journal of Urology Volume 166, Issue 5, November 2001, Pages 1738-1741

159. Ilana Roth-Deria, Tamar Green-Sadana and Gal Yadid. β-Endorphin and drug-induced reward and reinforcement. Progress in Neurobiology Volume 86, Issue 1, September 2008, Pages 1-21

160. David Bentonal and Rachael T Donohoe. The effects of nutrients on mood. Public Health Nutrition (1999), 2 : 403-409

161. Reginald Matejec, Axel Schulz, Heinz-W. Harbach,1 Holger Uhlich, Gunter Hempelmann, and Hansjörg Teschemacher2. Effects of tourniquet-induced ischemia on the release of proopiomelanocortin derivatives determined in peripheral blood plasma. J Appl Physiol 97: 1040-1045, 2004. First published May 14, 2004

162. R.I.M. Dunba. The social role of touch in humans and primates: Behavioural function and neurobiological mechanisms. Neuroscience & Biobehavioral Reviews Volume 34, Issue 2, 2010, Pages 260-268

163. J. T. Cheng, I. M. Liu, T. F. Tzeng, W. C. Chen, Hayakawa, T. Yamamoto. Release of β-Endorphin by Caffeic Acid to Lower Plasma Glucose in Streptozotocin-Induced Diabetic Rats. Horm Metab Res 2003; 35: 251-258.

Copyright Tanya Zilberter 2010.
Information: reflexo-therapy.com
Contact: panacea@reflexo-therapy.com

www.ingramcontent.com/pod-product-compliance
Lightning Source LLC
LaVergne TN
LVHW080006240125
802047LV00036B/1634